SUICIDE PREVENTION

SUICIDE PREVENTION

A Holistic Approach

Edited by

D. DE LEO

Department of Psychogeriatrics, Institute of Neurology and Psychiatry, University of Padua, Italy

A. SCHMIDTKE

Department of Clinical Psychology, Institute of Psychiatry, University of Würzburg, Germany

and

R. F. W. DIEKSTRA

Community Mental Health Center, Leiden and Municipal Health Authority, Rotterdam, The Netherlands

Springer-Science+Business Media, B.V.

Library of Congress Cataloging-in-Publication Data

Suicide prevention : a holistic approach / edited by D. De Leo and R.F.W. Diekstra.
p. cm.
Includes index.
ISBN 978-90-481-4815-8 ISBN 978-0-306-47210-7 (eBook)
DOI 10.1007/978-0-306-47210-7
1. Suicide--Congresses. 2. Suicidal behavior--Congresses. 3. Suicide--Prevention--Congresses. I. De Leo, Diego, 1951- II. Diekstra, R. F. W.
[DNLM: 1. Suicide--prevention & control--congresses. 2. Suicide--psychology--congresses. 3. Holistic Health--congresses. HV 6545 S948545 1997]
RC569.S93626 1997
362.28'7--dc21
DNLM/DLC
for Library of Congress 97-232

ISBN 978-90-481-4815-8

Printed on acid-free paper

Table of Contents

Preface

Suicide prevention is still sadly neglected today by governments and public health authorities, despite the fact that in several Western countries the phenomenon has become the first cause of death among the younger age groups, with a higher mortality rate than for road accidents. The World Bank and the World Health Organization deem that suicide causes at least 800,000 deaths per year throughout the globe and that the number of attempted suicides is probably ten times higher. It follows that the people involved each year in suicidal behaviour are several million in number, often leaving a long-term legacy of emotional, social and economic distress.

These data, nonetheless, have not so far managed to provide backing for preventive schemes of the same magnitude as the ones developed to tackle other public health problems of much lower epidemiological dimensions, such as AIDS.

Some years ago, the European Office of the World Health Organization set up a wide-scale multicentre study with a view to sensitizing national government authorities to the problem of suicide prevention, yet to date only a very limited number of countries have actually promoted national programs. Amongst these are Finland, Sweden, Norway and England. This apparently contrasts with the multitude of existing national associations for suicide prevention (although to be truthful, they are often composed of very few active members) or supranational associations, such as the International Association for Suicide Prevention (IASP), Befrienders International and the International Academy for Suicide Research (IASR), the latter group addressing in particular the promotion of high standard scientific study. "Man's only true philosophical problem", to quote Camus, is, however, so complex to prevent any attempt at comparison with other public-health issues, such as road accidents. Consequently, strategies to counteract the spread of the phenomenon must, by necessity, be highly integrated. In other words, suicide prevention cannot be limited to psychosocial, or by contrast, to biological considerations alone. Yet in spite of the many researchers in the field of suicidology, only a handful of culturally equipped research teams adopt a truly multidisciplinary orientation, with obvious repercussions on the development of the effective prevention measures the international community still awaits.

In June 1995, under the auspices of the International Association for Suicide Prevention, and sponsored by the World Health Organization and the

D. De Leo, A. Schmidtke and R.F.W. Diekstra (eds.), Suicide Prevention, ix–x.

European Economic Community, a congress was organized in Venice with the aim of promoting this very type of integrated approach to suicide prevention. This volume is derived from that event and contains some of the contributions which best express the need for a holistic viewpoint in suicide management: from the above-mentioned WHO multicentre study on parasuicide to the role of the media, from the special type of psychotherapeutic approach required to the most recent guidelines in pharmacological treatment, from a homage to the memory of Erwin Ringel (founder of IASP and undoubtedly one of the fathers of suicide prevention), to the presentation of specific national prevention schemes.

This book is dedicated to public health workers, doctors, psychologists and social workers, as well as voluntary staff and their organizations, and all those who make suicide prevention one of their primary interests.

D. De Leo
A. Schmidtke
R. F. W. Diekstra

1. Reflections on the State of Suicidology

RENE F. W. DIEKSTRA

I. Introduction

In 1995, at the time of the XVIII Biannual Conference of The International Association for Suicide Prevention (IASP) in Venice, it was exactly 35 years ago that Erwin Ringel took the initiative, together with a small group of colleagues, to establish IASP. Just as Vienna had once been the place in which the first conference on suicide ever took place (Friedman, 1967), namely in 1910, it also became the city in which half a century later, in 1960, IASP was born and its first conference was held. And just as the initiative to organize the 1910 meeting was taken by Alfred Adler, be it chaired by Freud, so the first IASP meeting in 1960 was initiated and chaired by an adlerian, for Erwin Ringel by training and attitude was first and foremost an adlerian psychiatrist. Certainly, we are not simply observing coincidence here. In a recent biography on Adler, entitled The Drive For Self (1995), psychologist Edward Hoffman noted that Alfred Adler among the great psychologists and psychiatrists of our era was the first and foremost advocate of the view that psychology should be brought to the people instead of the other way around. Adler, in contrast to Freud, was also an optimist: "he felt that every person has the potential to overcome the effects of a bad childhood and to become master of his or her own life". That, in the end, he considered to be the ultimate goal of psychological help, of mental health care: to empower people, to increase their sense of competence, their belief in self-efficacy as another outstanding psychologist, Albert Bandura, has expressed the same (Bandura, 1995). Educating people to help themselves, educating families to deal with the ebb and flow of conflicts in a healthy way, educating communities such as schools, companies and neighbourhoods to foster self-efficacy in their people was Adler's recipe for eradicating feelings of hopelessness/helplessness that form the breeding ground of demoralization, fatigue of life, depression and self destruction.

What has been achieved in this respect since the symposium of 1910? What has been achieved in this respect since the first IASP-meeting in 1960?

D. De Leo, A. Schmidtke and R.F.W. Diekstra (eds.), Suicide Prevention, 1–13.

What knowledge, what insights do we have that were not available then? What methods and techniques of prevention and treatment do we have at our disposal that were not available in those years? Often, voices can be heard both from within the community of suicidologists as well as from outside that we have been going around in circles, that basically our knowledge has not advanced and that our efficacy in preventing suicides has not grown over the course of this century. Quite a few outstanding suicidologists for that reason have after a number of years turned away from this field and redirected their intellectual and professional energy into other realms.

A. Progress in the Epidemiology of Suicidal Behavior

Are they right? Have we indeed made so little progress that Freud's concluding statement at the 1910 symposium still holds true:

> Gentlemen, I have the impression that, inspite of all the valuable material that has been brought before us in this discussion, we have not reached a decision on the problem that interests us. We are anxious above all to know how it becomes possible for the extraordinarily powerful life instinct to become overcome: whether this can only come about with the help of a disappointed libido or whether the ego can renounce its self-preservation for its own egoistic motives. It may be that we have failed to answer this psychological question because we have no adequate means of approaching it. (Friedman, 1967, p. 140)

The answer, I do believe, is both a no and a yes. No, the sceptics are wrong in that we do know much more today about the nature and magnitude of suicidal phenomena then we did a whole or half a century ago. While at the beginning of this century suicide was a phenomenon that aroused suspicion, silence, condemnation, moral indignation and could only be discussed by a small group of at the time still rather obscure psychoanalysts, at this century's end suicide has become a phenomenon that draws the attention of the public, of politicians, of health policy makers and even economists. And rightly so, because suicidologists have been able to demonstrate convincingly that what we have here is a health and social problem of the first degree, with a magnitude that is far greater than was supposed for the longer part of this century. This was clearly shown in the 1993 World Bank report (World Bank, 1993) where the best substantiated estimates of suicide mortality the world over were compared with estimates of mortality by other causes, such as motor vehicle accidents. Figure 1, which shows the global death toll by suicide compared to a number of other selected causes, makes a clear point.

Although the mortality by suicide is similar to the mortality by motor vehicle accidents and far greater than mortality through causes such as homicide/violence or war, the resources devoted to the prevention of suicide are only a fraction of those devoted to prevention of motor vehicle accidents. Of

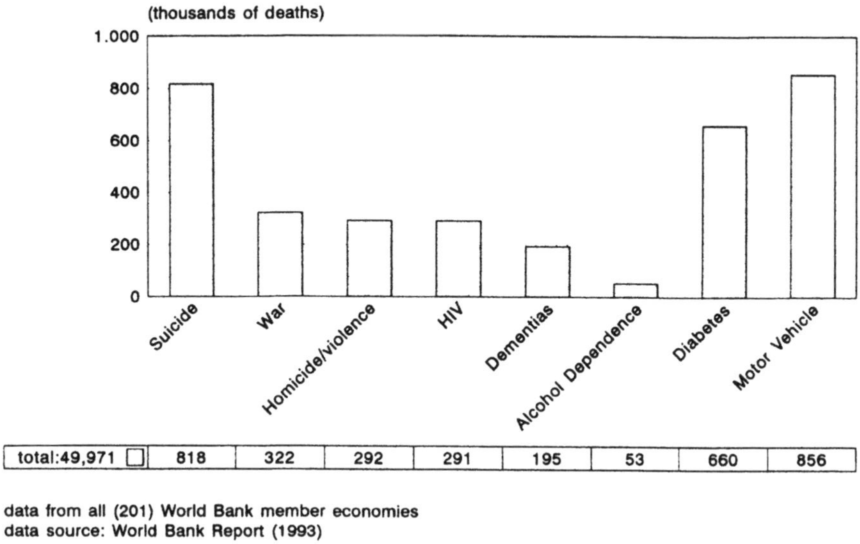

Figure 1. The burden of suicide in the world today, number of deaths by suicide vs. other selected causes, 1990.

course, in addition to deaths, motor verhicle accidents cause a much larger number of injuries, including a number of serious ones. This, however, cannot explain why much larger resources are devoted to road accident prevention than to suicide prevention. For non-fatal suicidal behaviour has to be considered, along with completed suicides, just as non-fatal accidents have to be considered with fatal ones. In addition to the number of suicidal deaths, it has for a long time been estimated that at least ten times as many persons make a non-fatal attempt to harm themselves, often serious enough to require medical attention and not infrequently resulting in irreversible disability. Such estimates were however based upon data of suicide attempts that led to hospital admission or contact with health agencies. It has been shown, however, that the majority of suicide attempts or attempts at deliberate self harm remain unknown to or unregistered by such agencies. Estimates are (see Figure 2) that this "tip of the iceberg" phenomenon implies that for every non-fatal suicidal act that leads to health care contacts there are at least three such acts that do not (CDC, 1991, Diekstra & Van de Loo, 1978).

But there is more to this activity of ever growing comprehensive data collection on the nature and magnitude of suicidal behaviour than meets the eye, and something of far greater importance than the data themselves. And that is the fact that suicidologists are becoming, be it hesitantly but certainly, guardians or "monitors" of mental health and social well-being in local and national communities and even the global community. By becoming more and more meticulous and complete in assembling information on the mor-

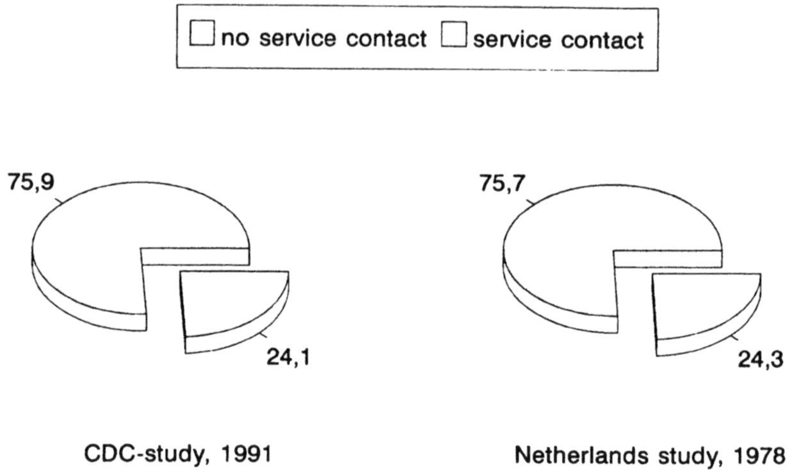

Student Risk Behavior Survey, 50 states, USA (CDC,1991)
Population Sample Survey, The Netherlands (Diekstra et al.,1978)

Figure 2. Parasuicide: the Tip of the Iceberg phenomenon.

tality by and the morbidity implicated in suicidal behaviour and more audacious in feeding that information back to the society, they do not only prevent denial and increase awareness, but they are to a certain extent also mobilising society itself in the war against premature death and mental ill-health. This is clearly demonstrated by the fact that a number of countries, such as Finland, The Netherlands, Norway and Sweden have in the course of the past decades launched national suicide prevention programmes.

But, we should ask at the same time, has the global community of suicidologists grown audacious enough in feeding back the information they assemble to society at large and to specific target groups in particular? The answer is clearly negative, for how else could it be then that a problem that only 20 years ago did not exist and that still is of far lesser magnitude, globally speaking, namely HIV-infection and AIDS, is at the forefront of the public stage while suicide still lingers backstage? Why is it that there is an AIDS Memorial Day, on which people all over the world come together in churches and other gathering places to commemorate the victims of the disease and light a candle for each one of them, and why it is that there is not a Suicide Memorial Day? Why is it that we have each and every year World AIDS Day, on which national and international organizations hand in hand with the media make people all over the world aware of the latest data and developments regarding this problem, and why is it that we have no World Suicide Prevention Day?

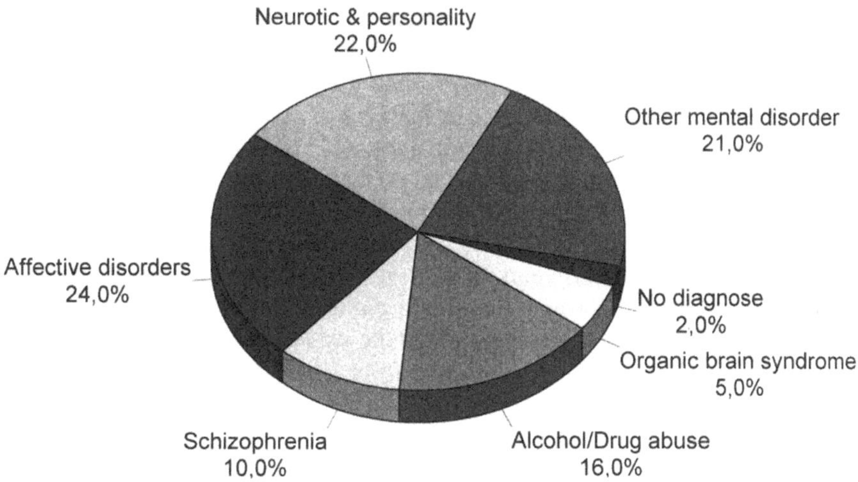

Figure 3. Suicide and mental disorder. WHO, 1993

Although, indeed, too many suicidologists still lack audacity, it is not just this factor that is to blame for these facts. As I will point out later, it is also the nature of the suicide phenomenon that stands in its own way.

B. Progress in Explaining Suicidal Behavior

But let me first address two other questions. The first one of which is this: are we better able to explain and predict suicidal behaviour than we were a century or half a century ago? This question is much more difficult to answer, and the answer will be different depending on where we look. It is for example questionable whether contemporary sociological explanations of suicide have significantly advanced beyond Emile Durkheim's theory of 1897 (Durkheim, 1897). Even most of the social risk factors emphasized by suicidologists today had already been identified or at least suggested by Durkheim and his students.

But it is unquestionable that we have made substantial progress with regard to the identification of psychological risk factors of suicidal behaviour and therewith with our potential for identifying high risk suicidal persons, at least in short term. However, identifying risk factors of suicidal behaviour is not the same as explaining that behaviour. Nor does it necessarily provide any insight into why the risk factor is a risk factor. Mental illness, for example, clearly is a risk factor, but a non-specific one since the increase in suicide risk related to its presence is not restricted to one particular disorder or category of disorders, such as depressive disorders, as Figure 3 (based on a review of the literature, see WHO, 1993) shows.

Generally speaking, we do not yet have a theory explaining a picture like the one presented in Figure 3. The same holds true for biological risk factors. We do know substantially more now than we did half a century ago with regard to biological correlates or markers of certain subtypes of suicidal persons. But the qualification "certain subtypes" is a significant one. We do not have a well-substantiated biological theory or a network of such theories of suicidal behaviour. And consequently we do not have a well-substantiated bio-psycho-social theory of suicide. Nor is it likely that we will have such a theory in another hundred years, if ever. It is even questionable whether we should strive for one. For again, as I will point out later, it is also the nature of the suicide phenomenon that stands in the way of such a theory.

C. Progress in Preventing Suicidal Behavior

And finally, what about our capacity to prevent suicidal behaviour? What advancements have we made in this respect? In terms of projects and programmes an awful lot. The number of people, lay people, volunteers, semi-professionals and professionals, around the world devoting their time, energy, affection and intellect in one way or another to those who are suicidal or in despair is simply immense, and in my view, one of the most impressive testimonies of Emmanuel Kant's and James Wilson's assertions (Wilson, 1993) that man indeed has an innate moral sense, a capacity for sympathy, a faculty for allowing him- or herself to be influenced by the experiences and feelings of fellow-human beings. If anything, suicides and suicidologists have mobilized communities indeed.

But, we must unavoidably ask, does it work, are those efforts effective, do they prevent preventive death and selfharm in significant and demonstrable ways? There is a growing number of "sceptics", even within the community of suicidologists, especially those with a biomedical view, who assert that the answer to this question has to be in the negative. Or at best, the verdict is: unproven with the exception of a very few spots here and there where the situation for very specific subgroups and very time consuming methods seems to be more positive, such as the work carried out by Marsha Linehan with female borderline patients (MacLeod et al. 1992) and even here its questionable whether what is prevented is death and not just the repetition of non-fatal behaviour. Some skeptics go even as far as stating that since suicidologists are not able to prevent suicide, to prevent death, they are now watering down their outcome-criteria to non-fatal selfharm and to even increased readiness to accept professional help (see Gunnellir & Frankel, 1994). Such criticism is, I believe, unfair or unjust for a number of reasons.

Imagine a person who is physically so ill that death is a very high probability. How effective is contemporary biomedical medicine in preventing death in such extreme cases? Now imagine a person who is psychologically so ill that death is a very high probability. How effective is psychiatry or

TABLE 1

Suicide in the industrialized world today

- While physical health status has improved, there has been a substantial increase in psychosocial disorders in youth since WW II in nearly all developed countries and of more recent date, in many developing countries as well.
- The evidence is particularly strong for crime/antisocial behaviour, alcohol and drug abuse, depression, and suicidal behaviour.
- For most of these disorders the trend is towards a convergence between rates for males and females (except for suicide per se).
- Similar increases have not been observed among elderly people.

psychology in preventing death, which is usually suicide, in such extreme cases? Is 'biomedical' medicine more effective in preventing deaths in the former case than psychiatry and psychology are in the latter? I seriously doubt it, but in any case there is no evidence to declare the one superior to the other. Even if we specify the areas of comparison between the biomedical and psychiatric-psychological approaches, the criticism is unjustified. Is, for example, biomedical medicine more effective in preventing AIDS or myocardial infarction than suicidology in preventing suicide attempts?

The ebb and flow of life expectancy and the emergence and disappearance or reduction of many if not most physical diseases is not related, for the larger part, to the workings of biomedical medicine. Likewise, the ebb and flow of mental problems and disorders is not controlled by the workings of psychiatry and psychology or other mental health disciplines. Nor will they ever be, for the larger part.

And then, we live in a century where socio-economic and cultural developments, at least in many parts of the world, have fostered physical health, while at the same time they appear to have had a deteriorating effect on mental health. In 1995 the Academia Europea published a study on time trends and causes of psychosocial disorders, including suicide, among young people in Europe covering the period of almost the whole of this century (Rutter & Smith, 1995). Some of the main conclusions of this milestone study are presented in Tables 1 and 2.

In other words, suicidology is rowing against the current and for that reason alone (if not for a number of other reasons such as the low base rate of suicide) it has been, and still is, almost impossible to prove the effects of preventive efforts on a population level.

But there is certainly also another reason for the lack of evidence on the efficacy of (preventive) interventions, one that has very much to do with the dominant approach within suicidology itself with regard to prevention and intervention. A recent meta-analysis of the 44 suicide intervention/prevention studies published in the period 1975–1994 and using a randomized control

TABLE 2

Types of intervention

	Curative	Preventive
Targeted at individual	Pharmacotherapy, ECT, psychotherapy counseling, extended follow-up contact, restriction of access to means, restriction for models, service provison/service improvement, training of service personnel I	Suicide education/life skills/health promotion programmes for the population II
Targeted at social network of group/community	Reconstruction of social and religious support system in bereaved community (refugees) III	? IV

I >> II >> >> >> III >> >> >> >> IV

design or an acceptable approximation thereof showed that these studies can appropriately be classified into the following four categories (Diekstra, 1996):

1. targeted at the individual and curative
2. targeted at group/community
3. targeted at individual and preventive curative, and
4. targeted at group/community and preventive

"Curative" refers to strategies that focus either on individuals or on groups/communities where one or more well-established risk factors of subsequent suicidal behaviour were known to be present. "Preventive" refers to strategies that focus on individuals or on groups/communities without prior assessment of the presence of one or more well-established risk factors (e.g. schoolbased programs addressing the total student population). "Targeted at individual" refers to strategies that through various methods or approaches seek to influence or modify affective, cognitive and/or behaviour patterns of individuals. "Targeted at group/community" refers to strategies that seek to influence or change the functioning of groups or communities as a whole. By far the majority of the 44 studies appeared to belong to category 1. A few studies could be classified within category 2, but only one study could be placed within category 3 and none in category 4 (see Table 2).

The meta-analysis, separately for suicide, parasuicide/attempted suicide and suicidal ideation, indicated that the efficacy of category 1 methods/ strategies, if one uses the magnitude of overall effect sizes as a criterion, is "not proven", although some methods can preliminary be labeled as "promising". As for the other categories, it was not possible at all to calculate effect sizes. In other words, (preventive) intervention studies in suicidology have almost exclusively focused on methods or strategies tailored at individual-curative level, which boils down to an essential problem-group approach, and have almost completely neglected the development and examination of methods/strategies that fall within the other three categories. Consequently suicidology has also failed to develop really comprehensive prevention programmes, that is to say programmes based on a well-substantiated methodology for the concerted implementation of methods/strategies from all of the categories. Until the time that such programmes are available it is highly unlikely that preventive efforts will be shown to take effects that are demonstrable at the general population or community level.

And finally, there is the nature of suicide itself, which in many ways forms an obstacle to its description, explanation and prevention. In the following I will try to elucidate why this is so.

II. The Nature of Suicide as an Obstacle to a Science of Suicide

It was around noon on a bright sunny day in august that 19-year-old Jürgen Peters climbed the ladder on the outside of the water tower in the German city of Kassel. By the time he reached the top, a number of people were already gathering at the foot of the tower, wondering and guessing what the young man was at.

It soon became clear that he intended to jump all the way down in an attempt to take his own life. Earlier that morning, Jürgen had been fired by his boss, a local garage owner, for whom he worked as an apprentice mechanic. The reason had been that, upon being asked to test drive a client's car, he instead had gone joy riding and in the process had severely damaged that car as well as two others.

Onlookers called the police, who in turn called the fire department for assistance. A fireladder was put out to the top of the tower, and one of the firemen tried to talk Jürgen out of his plan. Without success, however. Then a girl he had been dating and liked very much, was asked to talk to him. She climbed halfway up the fireladder, spoke to him through a megaphone for quite some time and succeeded in persuading him to give up his attempt. While Jürgen stepped from the roof of the water tower onto the fireladder and started his descent, a couple of young men watching the sequence of events, began to scream and yell at him: "Hey, you! Coward! You don't even have the guts to jump, do you?", and similar provocative remarks. One could see Jürgen hesitating at first, and then interrupting his descent. As he lingered there, just one meter or so below the top of the tower, an ominous silence descended over the scene. Then, all of a sudden, he started to move again but upwards, hopped on the top of the tower and, almost in one and the same movement, jumped off.

> During the few seconds that he fell along the length of the tower, the onlookers stood in petrified silence. That silence was brutally shattered by the dull sound of the body crashing on the earth. Panicked shouts and cries of anger and distress broke loose. They heralded an extremely painful period of mourning, depression, agression, disruption and hostility within the small community, that would last several years before it began to abate somewhat (Diekstra & De Leo, 1996).

However sad and shocking this event must have been at the time, and today still is for those who hear or read about it, in a number of ways it is also very revealing – instructive one would almost say if that word did not sound too cool and detached in this context. Jürgen Peters' suicide has been one of the very few cases in which the process of provoking and executing suicidal behaviour was actually "recorded before a live audience", so to speak. Nearly all cases of suicidal behaviour that scientists study, that practitioners deal with, and that the public reads or hears about, become known only after the fact. Even the most famous "suicidologists" of the past hundred years, whether their name is Durkheim, Halbwachs, Menninger or Shneidman, have never in their entire life been able to observe in actual fact the phenomenon to which they have devoted the larger part of their life, energy and intellect. In this respect suicidology, the scientific study of the nature, magnitude and preventability of suicidal behaviour, is to a large extent the pursuit of a secret. A dark secret, to many.

But that same fact also constitutes one of the main reasons of why suicide has fascinated mankind of all times. The quest for the revelation of the secret of suicide equals the quest for the Grail, the perennial chase by mankind for an absolute answer to the question of what makes life worth living, what justifies man's existence on this planet. This existential urge is what makes most suicidologists first and foremost 'preventionists'. They dream the dream of a world in which injustice can be amended, unhappiness mitigated, disease cured or palliated, to such extents that life, not death remains the final choice, and suicide a remnant of the past, a theoretical possibility at most. Time and again, however, they wake up to the brutal reality of a world which refuses to comply, which breaks the dream to pieces, which laughs all preventive efforts right in the face.

Why? Why has suicide always been so resistent to all those well-meant, well-planned and well-done cure and care programmes? What forces govern the ebb and flow of suicide mortality if it is not the availability and provision of qualified health and social care? The answer, for one thing, has to be found in the very nature of the phenomenon itself. Suicide is not a disease, unlike measles, smallpox, tuberculosis or Aids. Therefore, man's relationship with suicide was, is and will remain fundamentally different from his relationship with any of the other entities listed in the International Classification of Diseases. It is extremely rare, perhaps non-existent, for anybody in this world to plan, look forward or purposefully try to become affected by diseases such as measles, polio, tuberculosis or AIDS, or to ask another person for assistance in contracting such a disease.

Suicide is different. Although a large majority of people consider suicide to be highly undesirable, a considerable minority at one time or another during their lives welcome or seek out suicide, deliberately plan for its occurrence, sometimes with the assistance of others. Clearly, suicide is not a disease. Suicide is different. It is a behaviour, a *be-have-iour*, something that all humans *have* as an integral part of their *being*, of their existence. Suicide, in the most literal sense of the word, is an existential possibility, a *posse-ability*, something man has the ability to "pose", has the ability to do. We cannot contract every disease, even not in the unlikely event that we would want to do so. But we all can "contract" suicide. For that reason alone, if not for any other, we will never be able to eradicate suicide from the surface of our planet, unlike, for example, smallpox. This is not to say, though, that we are or will remain completely powerless in diminishing or removing a number of the motives for resorting to suicide, for choosing suicidal behaviour, and hence diminishing the actual toll of death and disability by suicide. But since suicide, like any other *behaviour*, can have an infinite number of motives, we will only be able to "annihilate" suicide completely at the expense of annihilating the very species we belong to, the homo sapiens sapiens. Making suicide impossible is making human life impossible, equals ripping manhood from humanity.

During that dark era we now call prehistory, there must once have been a nameless ancestor who was the first to discover that his of her life was not a necessity, not an unavoidable fact, not a prison; that he or she had the possibility to end it at will and by his or her own hand. The shock to the consciousness and the world-view of this unknown ancestor must have simply been gigantic. All of a sudden there was an alternative that would still be available once all other possibilities were exhausted; an escape route that would still be open once all other routes were blocked. Hardship, uncurable disease, torture, persecution, oppression, slavery, humiliation, from then on, they were no longer just fate.

Through the discovery of suicide, mankind discovered the maximum extent to which it could command its own destiny, could "free" itself. But the awareness of that freedom must also have filled it with awe, with fear. For from now on man's life no longer rested in the hands of God or the Gods only, but also in his own hands. But he was no God, and would he not offend God if he took over from him? Might just the thought of taking his life in and by his own hand not already evoke God's wrath? Not only upon himself but also upon those near and dear to him and therewith upon his community at large? Man's initial reaction to the, in the literal sense of the word, enormous possibility of suicide therefore had to be to throw it from him, forcefully and far: to reject, to forbid, to taboo suicide. A natural reaction almost, as Erich Fromm stated in his *Escape from Freedom*, for "the greatest freedom coincides with the greatest curse". Man cursed, had to curse, himself and his fellow beings for harbouring suicidal ideas and inclinations. Consequently he developed social and psychological mechanisms to transport such ideas and inclinations to places outside the walls of his consciousness and community: "extra muras". And whenever those defense mechanisms failed, which they were doomed to do every now and then, and suicide did occur, his response was fierce, cruel, and frightening, because he himself was extremely frightened by the act. Whether dead or alive, the suicidal person was, both literally and spiritually, cast out, removed from mankind, from the community of man, and from God's creation and mercy. He no longer was a "fellow" human being, no longer had any "fellows". Since he had irrevocably rejected life, others, his community, and God's creation, now he, in turn, was rejected by them, irrevocably. Condemnation, both in this life as well as in the hereafter, was his fate: perennial condemnation.

It took a redefinition of God, such as in early christianity, and later, after christianity had severely compromised the experience of God, a return to classicism and the early christian image of God, to make man, or at least some men, mitigate this attitude. Such as John Donne with his Biathanatos, which carries the full essence of this mitigation in its subtitle: *A Declaration of that Paradoxe, or Thesis that Self-Homicide is not so Naturally Sinne, that it may never be otherwise. Wherein the Nature and the extent of all those Lawes, which seeme to be violated by this Act are diligently surveyed.*

It took the removal of God self from the universe, to make some men push this mitigation into an apology of suicide, almost. Friederich Nietzsche, for example, not only declared God dead but also let his Zarathustra exclaim: "Einige sterben zu früh und vielen sterben zu spät. Und noch klingt fremd die Lehre: stirb zum rechten Zeit". (Some die too early and many die too late. And still the thesis resounds as a strange one: die on the right time!) Did it not, one may be inclined to ask, require the scientific pursuit of suicide, the emergence of suicidology, to pass the final sentence about which one of those three attitudes rests on the most solid, empirical ground? The answer is clearly negative. Science is as much an expression of the prevailing models of man and world as it is a factor of it. Despite a rapidly increasing body of knowledge, suicidologists had and still have as much a divided attitude towards the phenomenon they study as lay men have and have had for centuries. Suicide is not an unitary phenomenon, neither in terms of characteristics, nor in terms of pathways leading up to it, or in preventive possibilities and desirabilities. And it never will be.

It is my belief that the most fruitful, both socially as well as scientifically, point of departure of suicidology should be to respect this as a fact. For its task is not to point out and to substantiate what should be. Its task is first and foremost to point out what is: to describe, but not to prescribe. Its task is also to wonder and ponder on the possible whys of what is. And finally, its task is to indicate and demonstrate what can be done for people to change pathways, away from premature death and towards mature life. But it should be impartial enough not to forget, not to cast out, those fellow human beings for whom such a shift of pathway is not feasible, not possible, and sometimes perhaps undesirable. To acknowledge what one cannot do when one cannot do is not a sign of defeat, but of wisdom. And wisdom, as Herman Hesse put it in his Diesseits, is the ability to discern where knowledge ends and one is left with common sense and compassion as the only guiding lights.

I hope and I trust that the future "state of suicidology" will rest upon these four pillars, science, wisdom, compassion and common sense. No less.

References

Bandura A (editor). Self-efficacy in changing societies. Cambridge: Cambridge University Press, 1995.

CDC-Centers of Disease Control. Attempted suicide among high school students – United States 1990, leads from the Morbidity and Mortality Weekly Report. J. of The American Medical Association, 1991; 266: 14, 911.

Diekstra RFW, Van der Loo K. Attitudes towards suicide and incidence of suicidal behaviour in the general population. In: Winnick H, Miller L (editors). Aspects of suicide in modern civilization. Jerusalem: Jerusalem Academic Press, 1978; 79–85.

Diekstra RFW. A meta-analysis of suicide intervention and prevention studies: 1975–1994. Archives for Suicide Research, 1996; (accepted).

Diekstra RFW, De Leo D. The Anatomy of Suicide. A treatise on historical, social, psychological and biological aspects of suicidal behaviours and their preventability. Boston/Dordrecht: Kluwer Academic Publishers, 1996; *(in press)*.

Donne J. Biathanatos: *A Declaration of that Paradoxe, or Thesis that Self-Homicide is not so Naturally Sinne, that it may never be otherwise. Wherein the Nature and the extent of all those Lawes, which seeme to be violated by this Act are diligently surveyed.* (New Spelling Edition, 1982: New York: Garland), 1647.

Friedman P (editor). On Suicide: With particular reference to suicide among young students. New York: International Universities Press, 1967.

Gunell D, Frankel S. Prevention of Suicide: aspirations and evidence. British Medical Journal, 1994; 308: 1227–1233.

Hoffman F. The Drive for Self. New York: Addison-Wesley, 1995.

Macleod AK, Williams JG, Linehan MM. New developments in the treatment of suicidal behaviour. Behavioural Psychotherapy, 1992; 20, 30: 193–218.

Rutter M, Smith D editors. Psychosocial Disorders in Young People: Time Trends and Their Causes. Chichester: Wiley, 1995.

Wilson JQ. The Moral Sense. New York: The Free Press, 1993.

World Bank. Investing in Health. World Bank Report 1993. Oxford: Oxford University Press, 1993.

World Health Organisation. Guidelines for the primary prevention of mental, neurological and psychosocial disorders. 5. Suicide. Geneva: World Health Organisation, 1993; WHO/MNH/MND/93.25.

2. Suicidal Ideation and Suicide Attempts

The Role of Comorbidity with Depression, Anxiety Disorders, and Substance-use Disorder

T. BRONISCH and H.U. WITTCHEN

I. Introduction

Many studies in the past have investigated suicidal ideation, suicide attempts, and suicide in relationship to depressive disorders (Weissman, 1974; Roy, 1989; Ennis et al. 1989). Only very recently a few studies dealt with the potential importance of other disorders as well as comorbidity (the presence of more than one mental disorder, such as anxiety and depressive disorder) for the risk of suicidal behavior (Weissmann et al. 1989; Markowitz et al. 1989; Johnson et al. 1990; Petronis et al. 1990). Most of these studies have been primarily concerned with the relationship of suicidal ideation or suicide attempts and panic attacks or panic disorders with depression (Weissman et al. 1989; Markowitz et al. 1989; Johnson et al. 1990). The likelihood of suicide attempts for subjects with panic disorders or attacks has been reported to be similar to or greater than those associated with major depression (Markowitz et al. 1989).

Murphy and Wetzel (1990) estimated that the suicide rate for alcoholics in epidemiological studies is approximately 2.0–3.4%. Hawton et al. (1989) reported that alcoholics with prior histories of suicide attempts were at greater risk for additional attempts during the 10-year follow-up than nonalcoholics with previous suicide attempts. The following diagnoses have been found by Petronis et al. (1990) as risk factors for making a suicide attempt during a 1–2 year observation interval based on an analysis of data from the Epidemiologic Catchment Area (ECA) surveys in the United States: A current diagnosis of major depression, of alcohol dependency as well as any use of cocaine. All these above-mentioned studies suggest that major depression, panic attacks or disorder, alcoholism, and cocaine use are all diagnoses which contribute substantially to the risk of subjects making suicide attempts. However, the effect of comorbidity on the risk of making suicide attempts in regard to

D. De Leo, A. Schmidtke and R.F.W. Diekstra (eds.), Suicide Prevention, 15–26.

depression, substance-use disorder and panic attacks or disorders remains an open question.

We will report on the effect of comorbidity on rates of suicidal ideation and suicide attempts in an adult general population sample of the former West Germany (The Munich Follow-up Study, MFS; Wittchen and von Zerssen, 1988) using the Diagnostic Interview Schedule (DIS, Robins et al. 1981) as a standardized assessment instrument for DSM-III diagnoses as well as for the assessment of suicidal ideation and suicide attempts.

The following questions are addressed in this study:

1. How prevalent are suicidal ideation and suicide attempts in the general population?
2. How frequent are suicidal ideation and suicide attempts in different DSM-III disorders?
3. How frequent are suicide ideation and suicide attempts in comorbid cases?

II. Methods

A. Description of the Munich Follow-up Study

The Munich Follow-up Study (MFS) is a seven-year prospective and retrospective follow-up study of (a) a cohort of former psychiatric inpatients at the Max Planck Institute of Psychiatry in Munich (not reported here), and (b) a cohort of a general population sample of the adult population of what was West Germany, including West Berlin. The same evaluation methods were used for both samples (Wittchen, 1986, 1987, 1988; Wittchen et al. 1985). This paper exclusively describes results from the epidemiological sample. For details of the stratification of the epidemiological sample see Bronisch and Wittchen (1994).

The epidemiological sample was originally drawn in 1974, the year the phase-I investigation of the MFS took place. In this first phase 1952 of 2524 subjects (77.3%), randomly drawn from the general population, were interviewed for the first time (the refusal rate was 16.2%. 4.9% were either not available or did not complete the interview. 1.6% subjects had missing values in at least one of the scales used and thus were excluded from the further analysis).

Since a lot of other additional information collected proved to be very helpful for a prospective study, we decided to use this 1974 sample as a basis of our 1981 phase-II investigation for determining prevalence rates of mental disorders.

657 subjects were followed-up for 7 years (1974–1981) by monitoring health insurance records for the more detailed clinical follow-up investigation in 1981.

At the phase II investigation in 1981, 22 subjects (3.3%) had died, 97 (14.8%) refused the whole interview, and 37 (5.6%) refused parts of it. Of the 657 subjects, 501 (response rate 76.2%) could be interviewed in phase II. Of these 501 subjects, only 481 (73.5%) had a full data set and so only these subjects were used for this study. Prevalence rates for suicidal ideation and suicide attempts were weighted back to the original sample of 1366 from which the sample of 481 interviewed subjects was obtained. In addition, prevalence rates in % are given. The weighting was done in the following way: 1366 people were interviewed at the first time ("total population"), 1160 of them were low scorers, 206 were high scores. In the second interview 9 years later 454 of the 1160 low scorers were reanalyzed, as well as 127 of the 206 high scorers. We weighted the prevalence rates back to the "total population". This was done by weighting the low scorers with 1160/354 (3.28) and the high scorers with 206/127 (1.62). Thus the prevalence rates (%) for suicidal ideation and suicide attempts are weighted back to the original 1366 sample which formed the basis for our phase-II stratification.

Table I shows the sociodemographic characteristics of subjects who were interviewed in 1981.

Compared with the adult general population there is a slight preponderance of women. Most of the subjects were married, living in inner city areas and were employed.

B. Design and Study Instruments

In addition to other instruments (for details see Wittchen and von Zerssen, 1988), all subjects and patients were interviewed with the German version of the Diagnostic Interview Schedule (DIS, Robins et al. 1981; Wittchen and Rupp, 1981). This instrument allows the use of computer programs for scoring DIS information to produce diagnostic information according to DSM-III for lifetime, 6-month, and current diagnoses. In our study only lifetime diagnoses are considered. Anxiety (panic disorder, agoraphobia, simple and social phobia), depressive (major depression, single episode, recurrent, dysthymia, bipolar), substance-use disorders (alcohol abuse/dependence, drug abuse/dependence), somatization, schizophrenia, obsessive-compulsive disorder, psychosexual dysfunction, and eating disorders are included in the DIS used in the MFS, thus giving 16 possible DSM-III diagnoses. The DIS data on alcohol use are based on a modified alcohol section which included the results of a combined self-report questionnaire and observers's checklist (MALT; Feuerlein et al. 1979), instead of some of the original DIS questions (for description, see Bronisch and Wittchen, 1992) to ensure a high degree of comparability with the instruments used in phase I.

Although the DIS is designed for use by lay interviewers, only clinicians administered the DIS and all other instruments (Wittchen and von Zerssen 1988) in the MFS. These clinicians were either experienced physicians (more than 2 years of practical psychiatric training and experience after receiving their medical degree, n = 8) or clinical psychologists (n = 12). They were

TABLE 1

Sociodemographic characteristics of the general population sample

	General population sample (N=481)[1]	
	N	%
Sex		
Male	231	48.0
Female	250	52.0
Age		
25–34	80	16.6
35–44	169	35.1
45–54	130	27.0
55–63	102	21.2
Mean (s.d.)	44.85	(9.60)
Marital status		
Single	36	7.5
Married	388	80.7
Separated	7	1.5
Widowed	20	4.2
Divorced	30	6.2
Social class[2]		
I	18	3.8
II	73	15.3
III	214	45.0
IV	157	33.0
V	14	2.9

Five subjects in the general population did not have any data on social class

[2] Based on Hollingshead & Redlich (1958)

all trained in a 2-week video-assisted session in the use of the DIS, with the training material and a manual of instructions from the principal authors of the DIS. The interview training included further video-assisted practical experience under supervision throughout the whole study with the DIS as well as the other study instruments (Wittchen, 1984).

Table 2 shows the questions of the DIS concerning suicidal ideation and suicide attempts.

TABLE 2

DIS-Items with regard to suicidal ideation and suicide attempts (section "Depression")

D 88:	Thinking about death: Has there ever been a period of one week or more when you thought (Did you think) a lot about death – either your own, someone else's or death in general?
D 89:	Wish to die: Has there been a period of one week or more when you felt (Did you feel) like you wanted to die?
D 90:	Suicide ideas: Have you ever felt (Did you feel) so low you thought of committing suicide?
D 91:	Suicide attempts: Have you ever attempted (Did you attempt) suicide?

C. Statistical Analysis

Prevalence rates reported are weighted data. Weights used refer to the stratification described above. Adjusted odds ratios with 95% confidence intervals (95% CI) were calculated. The ratios indicate the strength of the association between the diagnostic groups and the suicide related variables. The statistical significance of the adjusted odds ratios can be judged from the confidence intervals (whether the interval excludes 1.0). A confidence interval that includes 1.0 indicates no statistical evidence for excess risk for the diagnostic group compared with no disorder. A confidence interval greater than and excluding 1.0 indicates increased risks for suicidal ideation and suicide attempts.

There was no adjustment done for the odds ratios. The calculation was done according to the normal formula:

	+	–	
+	a	b	
–	c	d	odds ratio = (a∗d)/(b∗c)

e.g. Major depression only, thinking about death:

	+	–	
+	352	14	
–	89	10	odds ratio = (352∗10)/(14∗89) = 2.8

TABLE 3

The weighted prevalence rates of suicidal ideation and suicide attempts in the MFS (N = 481)

DIS-Items	Total		Male		Female	
	N	Rates/100 (SD)	N	Rates/100 (SD)	N	Rates/100 (SD)
Thinking about death	99	18.5 (1.4)	26	11.0 (1.7)	73	25.5 (2.3)
Wish to die	33	5.3 (0.7)	6	2.5 (0,8)	27	7.6 (1.2)
Suicide ideas	83	14.7 (1.3)	30	12.5 (1,7)	53	16.3 (1.8)
Suicide attempts	18	4.1 (1.0)	5	2.2 (0.8)	13	4.1 (1.0)

III. Results

A. The Prevalence Rates of Suicide Ideation and Suicide Attempts in the MFS

Table 3 shows the weighted prevalence rates of suicidal ideation and suicide attempts in the MFS.

As can be seen from Table 3 the number of subjects "wishing to die" was lower than of subjects having "suicide ideas". However, suicide ideas are not dependent on the one week duration criterion as compared to the wishes to die (see Table 2). As expected the number of subjects with suicide attempts is considerably lower as compared to that of suicidal ideation.

B. The Frequency of Suicidal Ideation and Suicide Attempts and DSM-III Diagnoses

Table 4 shows the frequency of suicidal ideation and attempts across selected DSM-III diagnoses of the subjects. Only those diagnoses are considered where a sufficient number of cases with a DSM-III diagnosis were identified (at least five cases).

Subjects with no diagnosis have a lower frequency of suicidal ideation and suicide attempts as compared to the subjects with DSM-III diagnoses. The rates of suicide ideas and attempts are, however, rather similar for each of the DSM-III diagnoses considered.

Table 5 shows the comparison of subjects of the MFS with no DIS diagnosis, major depression without and with other DIS diagnoses (panic attacks, phobias, substance-use disorder) in regard to suicidal ideation and suicide attempts. Dysthymia was excluded as a condition never occuring in its pure form. Because no differences were found between alcohol abuse/dependency and medication abuse/dependency (since there were no cases of drug abuse/dependency we refer only to medication abuse/dependency), these

TABLE 4

DSM-III diagnoses and suicidal ideation and suicide attempts of subjects of tbe MFS (N =481)

DIS-Items	No DIS diagnosis		Major depression		Dysthymia		Panic attacks[a]		Phobias[b]		Alcohol abuse/ dependence		Drug abuse/ dependence	
Lifetime	N = 316		N = 54		N = 46		N = 53		N = 66		N = 67		N = 12	
	N	%	N	%	N	%	N	%	N	%	N	%	N	%
Thinking about death	46	15	30	56	20	44	19	36	25	38	13	19	4	33
Wish to die	6	2	20	37	16	35	11	21	10	15	8	12	2	17
Suicide ideas	25	8	37	69	30	65	23	43	24	36	14	21	4	33
Suicide attempt	6	2	8	15	4	9	7	13	6	9	5	8	2	17

[a] defined as meeting criteria A + B

[b] includes simple,social,and agoraphobias

[c] Medication abuse/dependence only

TABLE 5

Comparison of subjects of the MFS with no D1S diagnosis, to those with major depression without and with other DIS diagnoses (panic attacks, pbobias, substance use disorder) with regard to suicidal ideation and suicide attempts (N = 481)

DIS-Items Lifetime	No DIS diagnosis N=316		Major depression only N=24		Major depression Panic attacks N=23		Major depression Phobias N=21		Major depression + Substance use disorder N=9	
	N	%	N	%	N	%	N	%	N	%
Thinking about death	46	15	10	42	13	57	12	57	6	67
Wish to die	6	2	17	71	9	39	7	34	6	67
Suicide ideas	25	8	15	63	18	78	15	71	8	89
Suicide attempts	6	2	1	4	6	26	4	19	3	33

two diagnoses were combined to one category of substance-use disorder to increase the group size.

The table reveals markedly higher rates for pure depression in suicidal ideation but not in suicide attempts. The comparison between subjects with a pure major depression and subjects with a major depression additionally with either panic attacks or phobias or substance use disorder showed markedly higher rates of suicide attempts of the comorbid diagnostic groups. Unfortunately, there were not enough cases with pure panic attacks, phobias, and substance use disorder for more detailed comparisons.

C. The Comorbidity of DSM-III Diagnoses with Suicidal Ideation and Suicide Attempts

Table 6 shows the odds ratios of subjects of the MFS with any DSM-III diagnoses, major depression without and with other DIS diagnoses (panic attacks, phobias, substance use disorder) with regard to suicidal ideation and suicide attempts.

Cases with any DSM-III diagnoses show significantly higher odds ratios of suicidal ideation and suicide attempts than the subjects without a DSM-III diagnosis. The highest odds ratios were obtained by the subjects with a major depression with panic attacks and with substance use disorder, whereas subjects with a pure major depression did not display a significantly elevated odds ratio.

TABLE 6

Odds ratios of subjects of the MFS with any DIS diagnosis and with major depression without and with other DIS diagnoses (panic attacks, phobias, substance-use disorders) with regard to suicidal ideation and suicide attempts

DIS-Items Lifetime	No DIS diagnosis N=165		Major depression only N=24		Major depression Panic attacks N=23		Major depression Phobias N=21		Major depression + Substance-use disorder N=9	
	OR	95% CI	OR	95% CI	OR	95% CI	OR	95% CI	OR	95% CI
Thinking about death	2.7*	(1.7–4.3)	2.8*	(1.2–6.8)	5.4*	(2.3–12.7)	5.5*	(2.2–13.4)	7.8*	(1.9–31.8)
Wish to die	9.8*	(4.0–24.3)	6.5*	(2.5–17.1)	11.1*	(4.4–28.3)	8.0*	(3.0–21.5)	31.6*	(7.5–133.1)
Suicide ideas	6.1*	(3.7–10.3)	9.0*	(3.8–21.6)	20.7*	(7.4–57.7)	13.8*	(5.1–36.6	40.3*	(5.0–327.2)
Suicide attempts	3.9*	(1.5–10.7)	1.1	(0.1–8.5)	12.6*	(4.2–37.6)	7.2*	(2.1–24.2)	14.6*	(3.3–64.2)

* = significantly elevated

IV. Discussion

Our survey – the Munich Follow-up Study (MFS) – is based on a small, but nevertheless representative sample of West German households (Wittchen et al. 1992) using a standardized diagnostic instrument for DSM-III (DIS, Robins et al. 1981).

The rate of suicide attempts of 2.2 for males and 4.1 for females lifetime/100 is in the range of other studies in the USA, Canada, New Zealand, using the same diagnostic instrument (Weissman et al. 1993). Suicidal ideation (suicidal thoughts) are reported in 16% of the German population using self report scales (Korczak, 1988), very similar to our study (14.7% suicide ideas).

The result that only two subjects without a DIS/DSM-III diagnosis made a suicide attempt was expected. Epidemiological studies, using DSM-III or DSM-III-R diagnoses found a high percentage of depressive disorder, panic attacks and panic, as well as addictive disorders in subjects making suicide attempts (Weissman et al. 1989; Petronis et al. 1990).

Unexpected was the finding that cases with a pure major depression did not have a significantly higher odds ratio for suicide attempts than subjects with no DSM-III diagnosis. This is surprising since epidemiological studies assessing risk factors for subjects who committed a suicide attempt reported always a depressive disorder as the most important risk factor (Weissman, 1974; Ennis et al. 1989; Petronis et al. 1990). However, these studies did not address the issue of comorbidity, i.e. most cases might have had additionally another diagnosis besides a depressive disorder. The study of Petronis et al. (1990) included only active cases of a major depression within an observation period of one year and not, as we did, lifetime diagnoses. On the other hand, our sample size ist very small (only one "pure" major depressive made a suicide attempt), so that our result should be regarded as tentative. In contrast with the cases with a pure major depression all cases with a major depression and additionally a phobic disorder or a panic disorder or a substance use disorder showed very high odds ratios, especially those with panic disorders and substance-use disorders. Therefore, comorbidity seems to be a powerful risk factor for suicide attempts but not a pure depressive disorder itself. Unfortunately, our numbers of cases with panic attacks or panic disorder solely or with a substance use disorder solely were too small for the calculation of odds ratios. Finally, we cannot clarify with our study if suicidal ideation and suicide attempts coincided with the DSM-III diagnoses and if so with which DSM-III diagnoses.

Clearly our study can only document the critical role of comorbidity, but cannot answer the essential question of why and how comorbidity contributes to an elevated risk of suicidal ideation and suicide attempts. This is a task that might be solved best in prospective clinical follow-up studies that address the issue of comorbidity more specially (Wittchen, 1991; Wittchen et al. 1991). Methodologically however it should be taken into account that inpatients of a psychiatric hospital usually display extremely high rates of comorbidity with up to 90% (Wittchen and von Zerssen, 1988) of all patients. Thus outpatients with a lower degree of comorbidity might be more useful for this kind of research. Furthermore, personality disorders should also be assessed, since subjects and patients with personality disorders make suicide attempts more frequently (Ennis et al. 1989) and display a high degree of comorbidity with Axis-I DSM-III and DSM-III-R personality disorders (Alnaes and Torgersen, 1988; Fyer et al. 1988).

Acknowledgements

The authors thank Dipl.-Inf. Hildegard Pfister for the statistical analyses and her advice on an earlier draft of this paper. The data reported here are part of the Munich Follow-up Study (MFS), supported by the Robert Bosch Foundation. The MFS is a comprehensive 6–8 year follow-up investigation of former psychiatric inpatients and a general population sample. Principal investigators are Prof. Dr. H.U. Wittchen and Prof. Dr. D. von Zerssen. The following researchers and interviewers contributed significantly to this study: Sabine Dehmel, Rosmarie Debye-Eder, Toni Faltermaier, all Dipl.-Psych.; Heidemarie Hecht, Ph.D.; Christian Krieg, M.D.; Reinhold Laessle, Ph.D.; Wolfgang Maier-Diewald, Dipl.-Psych.; Hans-Ulrich Rupp, M.D.; Gert Semler, Dipl.-Psych.; Karin Werner-Eilert, Ph.D.; Monika Wueschner-Stockheim, Ph.D.; and Georg Wiedemann, M.D.. The clinical reexaminations reported in this paper were done by Michael Zaudig, M.D., and Gerhard Vogel, M.D.

References

Alnaes R, Torgersen S. The relationship between DSM-III symptom disorders (Axis I) and personality disorders (Axis II) in an outpatient population. Acta psychiatr scand 1988; 78, 485–492.

Bronisch T, Wittchen HU. Lifetime and 6-month diagnoses of abuse and dependence of alcohol in the Munich Follow-up Study. Eur Arch Psychiatry Neurol Sci, 1992; 241: 273–282.

Bronisch T, Wittchen HU. Suicidal ideation and suicide attempts: comorbidity with depression, anxiety disorders, and substance abuse disorder. Eur Arch Psychiatry Clin Neurosci, 1994; 244: 93–98.

Ennis J, Barnes A, Kennedy S, Trachtenberg DD. Depression in self-harm patients. Br J Psychiatry, 1989; 154: 41–47.

Feuerlein W, Kufner H, Ringer C, Antons K. Munchner Alkoholismustest (MALT) Manual, Beltz, Weinheim, 1979.

Fyer MR, Frances AJ, Sullivan T, Hurt SW, Clarkin J. Comorbidity of borderline personality disorder. Arch Gen Psychiatry, 1988; 45: 348–352.

Hawton K, Fagg J, McKeown SP. Alcoholism, alcohol and attempted suicide. Alcohol + Alcoholism, 1989; 24: 3–9.

Hollingshead AB, Redlich FC. Social class und mental illness. Wiley, New York, 1958.

Johnson J, Weissman MM, Klerman GL. Panic disorder, comorbidity, and suicide attempts. Arch Gen Psychiatry, 1990; 47: 805–808.

Korczak D. Estimation of suicidal behavior in representative epidemiologic studies. In: Moller HJ, Schmidtke A, Welz R (editors) Current issues of suicidology. Springer, Berlin Heidelberg New York, 1989.

Markowitz JS, Weissman MM, Ouellette R, Lish JD, Klerman GL. Quality of life in panic disorder. Arch Gen. Psychiatry, 1989; 46: 984–982.

Murphy GE, Wetzel RD. The lifetime risk of suicide in alcoholism. Arch Gen Psychiatry, 1990; 47: 383–392.

Petronis KR, Samuels JF, Moscicki EK, Anthony JC. An epidemiologic investigation of potential risk factors for suicide attempts. Soc Psychiatry Psychiatr Epidemiol, 1990; 25: 193–199.

Robins LN, Helzer JE, Croughan J, Ratcliff KR. National Institute of Mental Health Diagnostic Interview Schedule. Its history, characteristics, and validity. Arch Gen Psychiatry, 1981: 3381–389.

Roy A. Suicide. In: Kaplan HI, Sadock BJ (editors) Comprehensive Textbook of Psychiatry, 5th ed. Baltimore Md, Williams and Wilkins, 1989.

Weissman MM. The epidemiology of suicide attempts, 1960 to 1971. Arch Gen Psychiatry, 1974; 30: 737–746.

Weissman MM, Bland R, Joyce PR, Newman S, Wells EJ, Wittchen HU. Sex differences in rates of depression: Cross-national perspectives. J Aff Disorders, 1993; 29: 77–84.

Weissman MM, Klerman GL, Markowitz JS, Ouellette R. Suicidal ideation and suicide attempts in panic disorder and attacks. N Engl J Med, 1989; 321: 1209–1214.

Wittchen HU. The German version of the Diagnostic Interview Schedule (DIS, Version 2) – Reliability and results from a general population survey, report to the Division of Biometry and Epidemiology, NIMH, 1984.

Wittchen HU. Epidemiology of panic attacks and panic disorders. In: Hand I, Wittchen HU (editors) Panic and phobia. Springer, Berlin Heidelberg New York, 1986: 18–28.

Wittchen HU. Chronic difficulties and life events in the long term course of affective and anxiety disorders: Result from the Munich Follow-up Study. In: Angermeyer M (editor) From social class to social stress – new developments in psychiatric epidemiology. Springer, Berlin Heidelberg New York, 1987: 176–196.

Wittchen HU. Natural course and spontaneous remissions of untreated anxiety disorder: Results of the Munich Follow-up Study (MFS). In: Hand I, Wittchen HU (editors) Panic and Phobias 2. Springer, Berlin Heidelberg New York, 1988: 3–17.

Wittchen HU. Der Langzeitverlauf unbehandelter Angststörungen: Wie häufig sind Spontanremissionen? Verhaltenstherapie, 1991; 1: 273–282.

Wittchen HU, Essau CA, Krieg JC. Anxiety disorders: Similarities and differences of comorbidity in treated and untreated groups. Br J Psychiatry, 1991; 159: 23–33.

Wittchen HU, Essau CA, Zerssen D von, Krieg JC, Zaudig M. Lifetime and six-month prevalence of mental disorders in the Munich Follow-up Study. Eur Arch Psychiatry Clin Neurosci, 1992; 241: 247–258.

Wittchen HU, Rupp HU. Diagnostic Interview Schedule. German Version 2. Max Planck Institute for Psychiatry, Munich, 1981.

Wittchen HU, Semler G, Zerssen D von. A comparison of two diagnostic methods: clinical ICD diagnoses vs DSM-III and research diagnostic criteria using the Diagnostic Interview Schedule (version 2). Arch Gen Psychiatry, 1985; 42: 677–684.

Wittchen HU, Zerssen D von. Verläufe behandelter und unbehandelter Depressionen und Angststörungen. Springer, Berlin Heidelberg New York, 1988.

3. Depression, Hopelessness and Suicide Intent in Attempted Suicide

A Hospital-based Study of 201 Patients

A. T. DAVISON

Despite several recent statements that the reliable prediction of suicide is out of the question and may never be possible, clinicians continue to address the question of suicide risk as a matter of course in assessing the large number of patients who present to hospital following a suicide attempt. While it is widely recognised that the motives and meanings of attempted suicide patients are diverse and complex, it is generally assumed that attempted suicide patients with a high level of suicide intent resemble persons who commit suicide, both in respect of mental state and in personal and social characteristics.

Suicide intent, defined as the intensity of the wish to die at the time of the act, is a major factor to be considered in the evaluation of suicide risk, and remains an important area for study.

While the clinical states of depression and hopelessness in relation to suicide intent have been researched in some detail, little attention has been paid to other clinical and personal factors that may have a significant bearing on suicide intent in attempted suicide.

The inter-relationships among depression, hopelessness and suicide intent have been examined in several studies which are summarized in Tables 1a, 1b, 1c. Zero-order correlations have generally demonstrated significant positive correlations between depression and hopelessness and suicide intent. Partial correlations have generally shown that the relationship between depression and suicide intent disappears when hopelessness is controlled, but the reverse is not found. This has lead to claims that hopelessness is more closely related to suicide intent than is depression, and that hopelessness is the "missing link" between depression and suicide.

Prior to this study five other studies had examined other clinical and personal variables in relation to suicide intent. These studies are summarized in Table 2. It is apparent that the foci of studies have varied considerably and that findings are conflicting. Only three factors appear to be consistently associated with suicide intent, namely old age, presence of mental disorder (particularly depression) and living alone.

D. De Leo, A. Schmidtke and R.F.W. Diekstra (eds.), Suicide Prevention, 27–36.

TABLE 1a

Correlations between suicide intent, hopelessness and depression (Beck Depression Inventory)

Author (country)	Sample	Hopelessness	Depression	Hopelessness (depression controlled)	Depression (hopelessnes controlled)
Minkoff et al. 1973 (USA)	68 Attempted suicides	0.47***	0.26*	0.41***	−0.09 n.s.
Beck et al. 1975 (USA)	384 Attempted suicides	0.38***	0.30***	0.24***	0.06 n.s.
Dyer & Kreitman 1984 (UK)	120 Attempted suicides	0.50***	0.43***	0.32***	0.12 n.s.

*** $P < 0.001$; ** $P < 0.01$; * $P < 0.05$; n.s. not significant

TABLE 1b

Correlations between suicide intent, hopelessness and depression (Zung Self-rating Depression Scale)

Author (country)	Sample	Hopelessness	Depression	Hopelessness (depression controlled)	Depression (hopelessnes controlled)
Pokorny et al. 1975 (USA)	21 Depressed male attempted suicides	0.39 n.s.	0.52*	–	–
Wetzel 1976 (USA)	48 Attempted suicides	0.13 n.s.	0.21 n.s.	−0.04 n.s.	−0.17 n.s.

*** $P < 0.001$; ** $P < 0.01$; * $P < 0.05$; n.s. not significant

TABLE 1c

Correlations between suicide intent, hopelessness and depression ((a) Levine-Pilowsky Depression questionnaire, (b) Minnesota Multiphasic Personality Inventory)

Author (country)	Sample	Hopelessness	Depression	Hopelessness (depression controlled)	Depression (hopelessnes controlled)
(a) Goldney 1979 (Aust)	103 Female attempted suicides	0.45 ***	0.41***	0.26**	0.16 n.s.
(b) Wetzel et al. 1980 (USA)	73 Psychiatric inpatients (Intent scale for Ideators)	0.76*	0.36*	0.72*	–0.10 n.s.

*** $P < 0.001$; ** $P < 0.01$; * $P < 0.05$; n.s. not significant

TABLE 2

Socio-demographic and clinical variables related to suicide intent in attempted suicide

Lester, Beck & Trexler (1975; N=246)

- Older age
- Poor physical health
- Diagnosis of mental disorder

Pallis & Sainsbury (1976; N=151)

- Depression in preceding month

Pierce (1977; N=500)

- Older age
- Male sex
- Living alone
- Method other than self-poisoning
- Previous psychiatric treatment
- Previous attempted suicide

Power et al. (1985; N=80)

- Total life stress last 6/12
- GHQ "cases"

O'Brien et al. (1987; N=98)

- Older age
- Diagnosis major depression
- Living alone

• = significant association with suicide intent

Aims

This study aimed:

1. To further examine the relationship between degree of depression, hopelessness and suicide intent, using observer and self report ratings, and
2. To establish whether suicide intent is significantly associated with "high risk characteristics" of suicide.

The specific hypotheses to be tested were:

1. That there is a significant positive correlation between depression and hopelessness and suicide intent in attempted suicide.
2. That the relationship between depression and suicide intent is dependent on that between hopelessness and suicide intent.
3. That there is a significant association between suicide "high risk characteristics" and suicide intent in attempted suicide.

Method

The study was carried out at the Royal Adelaide Hospital over a three-year period.

Attempted suicide was defined as "a non-fatal act in which an individual deliberately causes self-injury or ingests a substance in excess of any prescribed or generally recognized therapeutic dosage". Patients with primary drug or alcohol intoxication were excluded.

Patients were eligible for this study if they were aged 18–65, spoke English and had no evidence of mental retardation or organic mental disorder. Subjects were interviewed as soon as possible after they had recovered from the adverse effects of self poisoning or self injury. This was generally possible within 48 hours of admission.

Following a comprehensive clinical interview, descriptive and demographic data and details of psychiatric history were recorded, and a DSM-III diagnosis was made. The degree of suicide intent was measured by the Suicide Intent Scale, which consists of 15 items, each rated on a 3 point scale. It allows both the circumstances of the suicide attempt and the self-reported suicide intent to be recorded and provides a total suicide intent score, a circumstances score and a self-report score. Depression was measured with the Zung Self Rating Depression Scale and the Hamilton Rating Scale for Depression. Hopelessness was assessed by the Hopelessness Scale. In completing the self report questionnaires, patients were asked to rate their mental state over the few days immediately prior to the suicide attempt. In rating the Hamilton Scale, an attempt was made to define the mental state over the same period of time.

Results

201 patients participated in the study. This comprised 80 males and 121 females, with mean age of 31 years.

A diagnosis of major affective disorder was made in 90 (45%) patients, adjustment disorder in 56 (28%) and other/no diagnoses in 55 (27%).

A comparison between the study group of 201 patients and the 352 non-study patients aged 18–65 admitted to hospital during the same period, showed no significant difference in mean age, sex ratio, marital status, employment status or method of attempt.

The mean Suicide Intent Scale score was 12.4 (SD 7.8), with mean circumstances score of 6.4 (SD 4.4) and mean self report score of 6.0 (SD 3.0). There was no significant difference in the correlations between self report and total score and circumstances and total score.

The mean Hopelessness Scale score was 12.6 (SD 5.3), the mean Zung SDS Index was 68.9 (SD 9.0) and mean Hamilton score was 18.2 (SD 6.5).

None of the measures was significantly related to sex. Correlated with age, only the Hamilton Score showed a significant but weak positive correlation ($r = 0.23$, $p < .01$).

Table 3 shows significant but weak positive Pearsons correlations of Hopelessness score, Zung score and Hamilton score with Suicide Intent score. Partial correlations between suicide intent and hopelessness, controlling for depression and between suicide intent and depression, controlling for hopelessness are shown. There were of a lesser magnitude than the zero-order correlations. With the Hamilton Scale both partial correlations remained significant at the $p < 0.05$ level. With the Zung Scale only hopelessness, controlling for depression, remained significant.

The Zung self report measure supports the finding of other authors, that the relationship between depression and suicide intent disappears when hopelessness is controlled. However this was not substantiated by the Hamilton observer rating scale.

The data were further examined by dividing Hopelessness, Zung-depression and Hamilton-depression scores about the median to define high and low depression, and high and low hopelessness groups. Two-way analyses of variance (suicide intent by hopelessness and depression) showed a significant main effect for hopelessness in both analyses ($p = 0.028$ and $p = 0.015$ respectively). Also, there was a significant interaction between hopelessness and Zung-depression, with members of higher score groups having higher suicide intent. This relationship was not found with Hamilton-depression.

Table 4 shows the relationship between socio-demographic and clinical variables which can define "high risk characteristics" of suicide and suicide intent. Using *t* test to compare mean SIS scores, only age >45, absence of alcohol intoxication and self-injury method of suicide attempt were found to be significantly associated with suicide intent. There was a striking lack of

TABLE 3

Correlations with Suicide Intent

Zero - order correlations		Partial correlations
Hopelessness	Zung-depression	Hopelessness (Zung -depression controlled)
$r = 0.19$ ($p < 0.01$)	$r = 0.16$ ($p < 0.05$)	$r = 0.13$ ($p = 0 < 0.05$)
		Zung-depression (Hopelessness controlled) $r = 0.08$ ($p = 0.14$)
	Hamilton-depression	Hopelessness
$r = 0.17$ ($p < 0.05$)	(Hamilton-depression controlled)	
		$r = 0.15$ ($p = 0 < 0.05$)
		Hamilton-depression (Hopelessness controlled) $r = 0.13$ ($p = 0 < 0.05$)

significant association of other items, particularly those related to affective disorder, alcohol abuse/dependence and previous suicide attempt, and suicide intent. These findings were reinforced by regression analysis which showed that absence of alcohol intoxication and self-injury method of attempt were the strongest predictors of high suicide intent.

Discussion

The study findings provide mixed support for the original hypotheses. Only weakly positive correlations between depression, hopelessness and suicide intent were demonstrated. With the Zung scale, the relationship between depression and suicide intent was found to be dependent on hopelessness. However, this was not the case with the Hamilton scale.

TABLE 4

Relationship of socio-demographic & clinical variables to Suicide Intent Score

Variable	N	Mean SIS	T-test p value
Sex			
Male	80	13.2	n.s.*
Female	121	11.9	
Age			
< 45	178	12.0	<0.05
> 45	23	15.6	
Marital Status			
Married/De facto	40	11.6	n.s.
Single/Separated/Widowed/ Divorced	161	12.6	
Employment			
Employed/Home duties/Student	108	12.2	n.s.
Unemployed/Pension	93	12.6	
Living circumstances			
Living alone	58	13.7	n.s.
Living with other(s)	143	11.8	
Method of suicide attempt			
Overdose	170	11.6	<0.01
Self injury	131	16.5	
Major physical illness			
Yes	41	12.0	n.s.
No	160	12.5	
Alcohol abuse/dependence			
Yes	40	10.7	n.s.
No	72	12.7	
Previous suicide attempt			
Yes	129	12.2	n.s.
No	72	12.7	

TABLE 4

Continued

Family history of suicide			
Yes	12	11.7	n.s.
No	189	12.6	
Major affective disorder (current)			
Yes	88	13.4	n.s.
No	113	11.6	
Alcohol intoxication			
Yes	46	8.2	<0.01
No	155	13.6	

*n.s. = not significant at 0.05 level

While several authors have extrapolated from attempted suicide studies and argued that hopelessness is the "missing link" between depression and suicide, this was not substantiated by this study. There are several methodological factors that may account for the conflicting findings:

1. variable definition of attempted suicide,
2. demographic differences and the size of study groups,
3. selection factors in admission to hospital, and further, to research projects,
4. the timing of assessments following the suicide attempt, and
5. the use of different measures of depression

Each of the studies summarized in Table 1 utilized self-report measures of depression. Perhaps the most important finding of this study is that an observer rating, which is generally considered to have greater validity in the assessment of depression than a self-report rating, has not substantiated the findings of these earlier studies. Clearly, further study of the complex relationship between mood, cognitive state and suicidal behaviour is needed.

While it is possible to systematically define and measure depression and hopelessness in a research setting, it is questionable that the two phenomena can be meaningfully separated in an experiential or clinical sense. In pragmatic terms it is hard to conceptualize severe hopelessness in the absence of severe depression, and vice versa. It may well be that to separate and study these concepts further in research studies will offer little practical assistance to clinicians in their attempt to understand and manage suicidal patients.

With regard to the suicide "high risk characteristics" examined, only three items have been found to be significantly associated with suicide intent, namely age >45, absence of alcohol intoxication and self-injury method of suicide attempt. This finding concurs with Hamdi et al. (1991), who reported a significant association between increasing age, psychosis and deliberate self-injury and high suicidal intent in attempted suicide. Pallis and Pierce (1979), in a 5-year follow-up study of 500 attempted suicides, found two of these factors to be highly correlated with eventual suicide. It appears therefore that these items deserve considerable weight in any evaluation of suicide risk factors in attempted suicide. At the same time it is important to be aware that research studies have not yet consistently identified definite predictors of suicide, even in this vulnerable population of suicide attempters.

References

Beck AT, Kovacs M, Weissman, A. Hopelessness and suicidal behaviour: an overview. Journal of the American Medical Association, 1975; 234: 1146–9.

Dyer JAT, Kreitman N. Hopelessness, depression and suicidal intent in parasuicide. British Journal of Psychiatry, 1984; 144: 127–133.

Goldney RD. Attempted suicide: correlates of lethality. MD Thesis, University of Adelaide, 1976.

Hamdi E, Amin Y, Mattar T. Clinical correlates of intent in attempted suicide. Acta Psychiatrica Scandinavica, 1991; 83: 406–11.

Lester D, Beck AT, Trexler L. Extrapolation from attempted suicides to completed suicides. Journal of Abnormal Psychology, 1975; 84, 5, 563–6.

Minkoff K, Bergman E, Beck AT, Beck R. Hopelessness, depression and attempted suicide. American Journal of Psychiatry, 1973, 130: 455–9.

O'Brien G, Holton AR, Hurren K, Watt L, Hassanyeh F. Deliberate self-harm – correlates of suicidal intent and severity of depression. Acta Psychiatrica Scandinavica, 1987; 75: 474–77.

Pallis DJ, Pierce DW. Recognizing the suicidal overdose. Journal of Royal Society of Medicine, 1979; 72: 1–7.

Pallis DJ, Sainsbury P. The value of assessing intent in attempted suicide. Psychological Medicine, 1976; 6: 487–92.

Pierce DW. Suicidal intent in self-injury. British Journal of Psychiatry, 1977; 130: 377–85.

Pokorny AD, Kaplan HB, Tsai SY. Hopelessness and attempted suicide: a reconsideration. American Journal of Psychiatry, 1975; 132, 9: 954–6.

Power KG, Cooke DH, Brooks DN. Life stress, medical lethality and suicidal intent. British Journal of Psychiatry, 1985; 147: 655–59.

Wetzel RD. Hopelessness, depression and suicide intent. Archives of General Psychiatry 1976; 33: 1069–73.

Wetzel RD, Margulies T, Davis R, Karam E. Hopelessness, depression and suicide intent. Journal of Clinical Psychiatry, 1980; 41, 5: 159–60.

4. Suicide Attempters who Attribute their Problems to Interpersonal Difficulties

KONRAD MICHEL and LADISLAV VALACH

In an earlier paper [1] we reported on the patients' own views about why they had attempted suicide. We made a distinction between reasons and causes which people give for their suicide attempt. Reasons (= motives) were most often expressed as "putting an end to an unbearable situation/state of mind", which we interpreted as the expression of the mental state of an emotional crisis. Causes (= problems), patients pointed out, were either of *inter*personal or *intra*personal nature. Interpersonal problems were e.g. problems with partner, parents, feelings of loneliness, etc., while intrapersonal problems were mental illness, addiction and physical illness. These results in our opinion are consistent with Shneidman's view [2] which states that the common purpose of suicide is to seek a solution and that there is a consistency in suicidal behaviour with lifelong patterns. Maris [3] put forward the concept of pathways to suicide (or suicide career) and pointed out that suicide can largely be seen as rational, problem-solving behaviour by people with similar life histories and that most suicide completers use self-destructive behaviour as a means of escape from a long accumulation of painful life experiences (or to aggress against another human being).

It is therefore important to understand a patient's suicidal behaviour in the context of his life development. Such an understanding, however, has largely to rely on the patient's own views about the causes of the suicide attempt. There is an association between the patient's understanding of the problem and age [4]. Younger patients tend to attribute the cause of a suicide attempt to relationship problems while older patients are more likely to attribute the cause to an illness such as depression. Others [5, 6] point out that the suicide of young adults is different from that of older adults. These findings may be understood as the expression of a development throughout a person's life, leading to an internalization of the problems and to increasing resignation.

In order to better understand the clinical relevance of the patients' views we wanted to know whether the suicide attempters attributing their problems to interpersonal difficulties differ from those ascribing them to intrapersonal (or health) issues. Obviously, the patients' retrospective accounts of the causes

D. De Leo, A. Schmidtke and R.F.W. Diekstra (eds.), Suicide Prevention, 37–44.

of their suicide attempts are subjective. On the other hand, any therapeutic interventions must take the subjective theory of a patient into consideration.

I. Method

A. Study Population

We interviewed a sample of suicide attempters which had been recorded in the monitoring project of the WHO/EURO Multicentre Study on Parasuicide. For the definition of parasuicide used in this study see Platt et al. 1992 [7]. Data on patients were collected from all general and psychiatric hospital units within a geographically defined catchment area comprising a population of 301 000.

Hundred and nineteen out of 327 suicide attempters aged 15+ subsequently admitted to hospital were asked by a member of the research team to participate in a structured interview. The other patients could not be contacted because they were discharged before they were reported to the research team or because of other reasons such as language problems. A few could not be interviewed because of a disturbed state of mind. Out of the 119 patients a further 53 refused, leaving altogether 66 suicide attempters with a completed interview, representing 20% of all known suicide attempters in the catchment area. A comparison with the original 327 persons showed no significant differences in age and sex distribution, civil status, living situation and employment status. The only difference found was in the level of education: the sample had a higher level of schooling ($Chi^2 = 17{,}58$, d.f. = 3, $p < 0.001$).

B. Interviewing Instruments

All 66 patients were interviewed with the European Parasuicide Study Interview Schedule (EPSIS, 8), a comprehensive interview schedule comprising several semi-structured questionnaires and self-rating scales. This interview schedule had been developed for the WHO/EURO Multicentre Study on Parasuicide and had been carefully translated and backtranslated from English into the different languages. All the interviewers had been trained in the correct use of this interview. The "Problem Check List" is a self-administered list of eleven possible problems, and the patients have to indicate if these problems had either had no influence at all on what they had done, or a minor or a major influence (see Table 1).

The 66 patients were divided into two groups: group 1 included patients (n=47) who on the Problem Check List indicated that one or several of the interpersonal problems presented to them had a major influence on what they had done. Group 2 consisted of the patients (n=19) who did not choose an interpersonal item as of major influence, i.e. who attributed most importance to problems concerning their physical or mental health. Unemployment

TABLE 1

Problem checklist[1]. Interpersonal problems are marked with an asterix

		Influence on what you did
1.	Problems with partner*	No / Minor / Major
2.	Problems with parents*	No / Minor / Major
3.	Problems with children*	No / Minor / Major
4.	Feelings of loneliness*	No / Minor / Major
5.	Problems with making or maintaining friendships*	No / Minor / Major
6.	Rejection by lover*	No / Minor / Major
7.	Physical illness or disability	No / Minor / Major
8.	Mental illness and psychiatric symptoms	No / Minor / Major
9.	Unemployment	No / Minor / Major
10.	Addiction	No / Minor / Major

[1)] From: European Parasuicide Study Interview Scheme (AJFM Kerkhof et al. 1990)

and addiction were both included under the intrapersonal problems. In the comparison of the two groups we were interested (1) in the social characteristics of the patients, (2) in quasi objective biographical information (e.g. parent's divorce), and (3) in normative assessments such as psychiatric diagnosis the Beck Depression Inventory [9], the Hopelessness Scale [10], and the Rosenberg self-esteem scale [11].

II. Results

A. Factor Analysis

A factor analysis based on the 10 items of the problem check list (see Table 1), without stipulation of the reasons for extraction, was performed with the data of all the patients (principal component analysis and Varimax rotation to simple structure after Kaiser Normalization). A four-factor solution was meaningful and explained 59.5% of the variance. The factor accounting for the largest part of variance (17.7%) is characterized by the item "rejection by lover" as interpersonal problem on one hand and "mental and psychiatric symptoms" as intrapersonal problem on the other hand. Furthermore, the structure of the correlation matrix of the reasons for suicide attempts shows a clear division between interpersonal issues on one hand and illness (mental or physical) on the other hand.

TABLE 2

Social characteristics

	Group 1 (N=47)	Group 2 (N=19)	p
Mean age	35.9	45.3	0.061 (t-test)
Gender			
male	40.4%	52.6%	n.s. (Chi^2test)
female	59.6%	47.4%	n.s.
Marital status			
single	46.8%	36.8%	n.s.
married	27.7%	31.6%	n.s.
divorced/sep.	21.3%	26.3%	n.s.
Education			
low	29.8%	26.3%	n.s.
middle	53.2%	47.4%	n.s.
high	17.0%	26.3%	n.s.
Living situation			
alone	25.5%	42.1%	n.s.
with partner and/or children	40.4%	26.3%	n.s.
other	34.0%	26.3%	n.s.

B. Social Characteristics of Patients

Table 2 shows that the two groups did not differ in the social characteristics. Although the mean age of the group 2 was higher (35.9 years vs 45.3 years), the difference was not significant ($t=1.96$, $p>0.05$).

C. Psychiatric Diagnosis

In group 1 a diagnosis of adjustment disorder (DSM III-R, 12) was made more frequently than in group 2 (36.2% vs 5.3%, Chi-square = 8.00, $p = 0.005$). Other frequent diagnoses in group 1 were: substance abuse, affective disorder, personality disorder, but there was no statistically significant difference between the two groups.

D. Scales

The groups did not differ in the Beck Depression Inventory (mean BDI 20.2, sd = 11.0, for group 1 and 15.3, sd = 10.2, for group 2; $t = -1.72$, $df = 35.5$, $p = 0.094$), or the Beck Hopelessness Scale (mean HS 10.1 and 8.5 respectively; $t = -1.02$, $df = 30.18$, $p = 0.32$). There was a significant difference, however, in the scores of the Rosenberg self-esteem scale (mean 3.34 in group 1 and 2.05 in group 2, respectively; Mann-Whitney U-test: $U = 260.0$, $p = 0.007$). Group 1 had a lower self-esteem, expressing high levels of dissatisfaction with and a lack of respect for themselves (e.g scale 5: "I wish I could have more respect for myself"; 87% yes answers in group 1, and 58% yes answers in group 2, respectively, Chi-square = 6.45, $p = 0.009$).

E. Indicators of Quality of Past Relationships

See Table 3. The main finding is that group 1 did not differ in the number of unsuccessful marriages but they did differ in that more had married again. It could be inferred that secure social relationships are important for them. This is also indicated by the fact that more of them live with others than those from group 2.

F. Help-Seeking Behaviour – Contact with GPs Prior to Attempt

Unlike the patients in group 2, the majority of patients with interpersonal reasons indicated that they had suicidal ideation during the time of their last visit to GP (53.5% vs 18.8%; Chi-square = 6.12, $df = 1$, $p = 0.013$), and just about half of these attempted to talk to their physician about it, while in group 2 only one sixth said they had done so.

III. Discussion

Suicide attempters had to indicate on a problem check list which problems had had an influence on their decision to attempt suicide. The items stipulated either interpersonal or intrapersonal problems. A factor analysis suggested that a distinction between interpersonal and intrapersonal problems is meaningful. Interpersonal items were chosen as having had a major influence by a majority of patients. We cannot exclude that this could partly be due to an error of sampling: It could be argued that the socially oriented patients are more willing to participate, which might bias the results. The relatively small sample of suicide attempters that could be interviewed is one of the limitations of this study. It will be necessary to repeat such an investigation with a larger number of patients.

The two groups did not differ in their social demographic characteristics. Patients in group 1 tended to be younger – which would support a devel-

TABLE 3

Indicators of quality of past relationships

	Group 1 (N=47)	Group 2 (N=19)	p (Chi^2-test)
Parents divorced	29.8%	5.3%	0.017
in childhood	17.0%	5.3%	n.s.
later in life	13.6%	0.0%	0.041
Parental relationship problems	72.3%	31.6%	0.002
in childhood	62.5%	31.6%	0.013
later in life	44.2%	16.7%	0.034
Rows with partner			
last year	38.2%	0.0%	0.001
before	32.4%	28.6%	n.s.
Those who were married:			
Divorced	48.0%	25.0%	n.s.
Married once	66.7%	100.0%	
Married twice	33.3%	0.0%	0.008
Living with others	62.5%	27.3%	0.041

opmental view of suicidal behaviour. Patients in group 1 indicated more difficulties in their past social relationships, e.g. their parents divorced more often. Patients themselves re-married more often, and reported serious relationship problems in the past year. It is unlikely that these answers are due to selective memorizing because they are "hard" biographical data. Thus, the patients' subjective attributions are supported by more objective data. The results in our view support the concept of lifelong patterns. The results of the Rosenberg self-esteem scale suggest that group 1 patients had a lower self-esteem and a tendency to blame themselves.

Consistent with these findings is that group 1 was more often given the diagnosis of adjustment disorder; we can infer that these patients rather have psychological problems while patients in group 2 more often suffer from psychiatric symptoms.

For the helper it is important to know that more than half of group 1 indicate that they had suicidal ideation at the time of the last contact with the GP. Although about 40% of the patients of the group 1 named psychological issues as a reason for their visit to a doctor, only about half of those patients managed to communicate their intentions to their practitioner. As is well known from other studies [13, 14, 15], there is a discrepancy between the frequency of suicidal thoughts and the frequency of the communication of these thoughts.

The results lead one to hypothesize about the different concepts of a "suicide career" [3] which are expressed by these two groups. A developmental model of suicidal behaviour encompasses several categories of career (or life development) failures (occupational failures, relationship failures, personal integrity failures). The suicide attempts in group 1, which indicated interpersonal problems, may be conceptualized as part of the social or relationship career, while suicide attempts in group 2 appear more as part of the "self or identity building career". In a developmental systems theoretic model psychological function and dysfunction are not primarily seen as properties of individuals but as properties of the complex interaction of person and environment.

To conclude, we found that the patients' subjective attribution is supported by objective biographical data. Furthermore, the results support the concept of lifelong patterns and of the importance of life histories in the understanding of suicidal behaviour. Suicide attempters who mention interpersonal problems as a reason for their act often have experienced relationship problems throughout their lives. We suggest that their suicide attempt should be seen in the context of their "relationship career". We might expect that if a potential helper addressed these interpersonal difficulties, these patients should be able to talk about it (because they appear to be aware of them). It therefore seems important to us that after a suicide attempt helpers search for interpersonal problems not only in the present but also in the past. The question remains what would be needed to improve the communication between patient and health professional. This, in our view, would be an important area of further research.

References

1. Michel K, Valach L, Waeber V. Understanding deliberate self-harm: The patients' views. Crisis, 1994; 15: 172–178.
2. Shneidman S. The commonalities of suicide across the life span. In: Leenaars AA (editor): Life perspectives of suicide, Plenum Press, New York, 1991.
3. Maris RW. Pathways to suicide: A survey of self-destructive behaviours. Johns Hopkins University Press, Baltimore, 1981.
4. Michel K, Jaeggi Ch, Sturzenegger M, Valach L. Charakteristika junger Suizidversucher und ihre Bedeutung für den Helfer. Psychother Psychosom med Psychol, 1994; 44: 1–6.
5. Leenaars AA. Suicide in the young adult. In: Leenaars AA (editor): Life perspectives of suicide, Plenum Press, New York, 1991.
6. Michel K. Suicide in Young People is different. Crisis, 1988; 912: 135–145.
7. Platt S, Bille-Brahe U, Kerkhof A, Schmidtke A, Bjerke T, Crepet P, De Leo D. Haring C, Lönnqvist J, Michel K, Philippe A, Pommereau X, Querejeta I, Salander-Renberg E, Temesvary B, Wasserman D, Sampaio Faria J. Parasuicide in Europe: the WHO/EURO

multicentre study on parasuicide. I. Introduction and preliminary analysis for 1989. Acta Psychiatr Scand, 1992; 85: 97–104.
8. Kerkhof A, Bernasco W, Bille-Brahe U, Schmidtke A. European Parasuicide Study Interview Schedule (EPSIS I). In: Bille-Brahe U et al.: Facts and Figures. WHO EUR/ICP/PSF 018, Copenhagen, 1993.
9. Beck AT, Ward CH, Mendelson M, et al. An inventory for measuring depression. Arch Gen Psychiatry, 1961; 4: 561–571.
10. Beck AT, Weissman A, Lester D, Trexler L. The measurement of pessimism: the hopelessness scale. J Consult Clin Psychol, 1974; 42, 6: 289–96.
11. Rosenberg M. Society and the adolescent self-image. Princeton University Press, Princeton, 1965.
12. American Psychiatric Association. DSMIII-R; Diagnostic and statistical manual of mental disorders (3rd edition). American Psychiatric Association, Washington DC, 1987.
13. Murphy GE. The physician's responsibility for suicide. I. An error of commission; II. Errors of omission. Ann Intern Med, 1975; 82: 305–309.
14. Michel K. Suizide und Suizidversuche: Könnte der Arzt mehr tun? Schweiz Med Wochenschr, 1986; 116: 770–774.
15. Choquet M, Menk H. Suicidal thoughts during early adolescence: prevalence, associated troubles and help-seeking behavior. Acta Psychiatr Scand, 1989; 81: 170–177.

5. Suicide among Psychiatric Hospital Inpatients

ALEC ROY and RONALD DRAPER

In 1982 we reported a matched controlled study of 90 psychiatric patients who committed suicide (Roy, 1982). Significantly more of the suicide victims suffered from chronic schizophrenia or affective disorder and had made a previous suicide attempt. Also, significantly more of the suicide victims were unmarried, unemployed and living alone. However, only 15 of those 90 suicide victims were psychiatric inpatients – too small a number in which to examine for risk factors for suicide among psychiatric inpatients.

We now report a matched controlled study of 37 patients of an Ontario Provincial Psychiatric Hospital who committed suicide while inpatients. The hypothesis tested was that there would be significant differences between the inpatient suicide victims and matched inpatient controls, which might suggest possible risk factors for suicide among psychiatric inpatients.

I. Subjects and Methods

Thirty-seven patients of Brockville Psychiatric Hospital, an Ontario Provincial Hospital, were known by audit process to have committed suicide between 1970 and 1991 while psychiatric inpatients. Each was matched for sex and age (within five years) with the next psychiatric patient admitted to the same hospital to constitute a control group. All of these charts, which included reports from other psychiatric and general hospitals, were examined and data extracted.

The suicide and control groups were compared for potential social and psychiatric risk factors. The social variables examined included marital status and whether living alone or unemployed at the time of the last admission. The psychiatric variables examined included the main psychiatric diagnosis made by the psychiatrist using the International Classification of Diseases (ICD), a history of a past suicide attempt, number of previous psychiatric admissions, whether ever treated for depression with antidepressants or electroconvulsive therapy (ECT) and whether there was a history of alcohol or drug abuse. In the last psychiatric admission five variables were recorded: age at suicide, whether admitted voluntarily or involuntarily, length of hospital stay up to suicide, site and method of suicide. The suicide and control groups were not

D. De Leo, A. Schmidtke and R.F.W. Diekstra (eds.), Suicide Prevention, 45–51.

matched for psychiatric diagnosis so as not to obscure potential diagnostic differences between the groups. We also examined the annual admission statistics as a check that the matched control group was representative of the total non-suicide hospital population.

In the statistical analysis chi square and Fisher's Exact tests two-tailed were used.

II. Results

Twenty-one of the suicide victims were single, 5 married and 11 divorced, separated or widowed. Their mean age at suicide was 38.9 ± 14.6 years. The only significant sex difference was in relation to marital status: more of the male than the female suicide victims were single (18 of 24 vs. 3 of 13, Fisher's Exact, $P < 0.005$) while more of the female suicide victims were divorced, separated or widowed (7 of 13 vs. 4 of 24, Fisher's Exact, $P < 0.03$).

Table 1 shows the data about sex distribution, length of last admission, site and method of suicide. Only one patient committed suicide within a week of admission while 12 of the suicide victims had been in hospital over a year. Only six patients committed suicide on the ward. The only significant sex differences were that more of the male than the female suicide victims committed suicide in front of a train (7 of 24 vs. 0 of 13, $\chi^2 = 3.84$, dfl, $P < 0.05$) while more of the females committed suicide by drowning (8 of 13 vs. 5 of 24, Fisher's Exact, $P < 0.03$).

Table 2 shows the primary diagnoses for the suicide victims and their inpatient controls. Significantly more of the suicide victims than controls had a diagnosis of schizophrenia ($\chi^2 = 12.3$, dfl, $P < 0.0005$). There was also a trend for a significant difference for epilepsy ($\chi^2 = 3.12$, dfl, $P < 0.077$). Significantly more of the controls had diagnoses of personality disorder ($\chi^2 = 8.6$, dfl, $P < 0.003$) and alcoholism ($\chi^2 = 4.2$, dfl, $P < 0.04$).

Table 3 shows the clinical data for the suicide victims and matched controls. The mean age of the controls was 36.7 ± 13.1 years. Significantly more of the suicide victims had made a past suicide attempt ($\chi^2 = 5.4$, dfl, $P < 0.02$), were involuntary at their last admission ($\chi^2 = 5.5$, dfl, $P < 0.02$) and lived alone ($\chi^2 = 5.5$, dfl, $P < 0.02$).

As the majority of the suicide victims were schizophrenic, we lastly compared only the schizophrenic suicide victims (N = 28) and schizophrenic controls (N = 13). In this comparison the only significant difference was that more of the schizophrenic suicide victims than schizophrenic controls had made a past suicide attempt ($\chi^2 = 7.3$, dfl, $P < 0.007$). There was a trend for more of the schizophrenic suicide victims to be living alone before their last admission ($\chi^2 = 2.45$, dfl, $P < 0.12$).

TABLE 1

Characteristics of Suicide

	No. of suicides N = 37
Length of Hospital Stay up to Suicide	
Less than a week	1
1 week – 1 month	5
Months 2,3	6
Months 4,5,6	6
Months 7–12	7
Over 1 year	12
Site of Suicide	
On the ward	6
Elsewhere in hospital	3
At home	2
Away from home and hospital	26
Method of Suicide	
Railway	7
Hanging	9
Run over	2
Overdose	3
Gun	2
Jump	1
Drown	13
Sex of Suicides	
Male	24
Female	13

III. Discussion

We carried out a matched controlled study of suicide among psychiatric inpatients at one Ontario Provincial Psychiatric Hospital. We found that 64.9% of the suicide victims were male, the mean age was 38.9 years, 32.4% had been in hospital over a year, and significantly more of the suicide victims than controls suffered from schizophrenia (75.7 vs. 35.1%), had made a previous suicide attempt (62.2 vs. 35.1%), were involuntary at their last admission (70.3 vs. 43.2%), and lived alone (70.3 vs. 43.2%).

The most frequent diagnosis in this Canadian series was schizophrenia. Similarly among the 97 and 71 state hospital inpatient suicides of Sletten

TABLE 2

Comparison of suicide victims and controls for psychiatric diagnosis

Psychiatric Diagnosis	No. of suicides N = 37	No. of controls N = 37	Significance
Schizophrenia	28	12	$\chi^2 = 12.3$ $P < 0.0005$
Affective Disorder	5	7	NS
Personality Disorder	1	10	$\chi^2 = 8.6$ $P < 0.003$
Epilepsy	3	0	$\chi^2 = 3.12$, $P < 0.077$
Alcoholism	0	4	$\chi^2 = 4.2$, $P < 0.04$
Adjustment Disorder	0	1	NS
Mental Retardation	0	2	NS

TABLE 3

Comparison of suicide victims and inpatient controls

	Suicide Victims N=37	Controls N=37	Significance
Past suicide attempt	23	13	$\chi^2 = 5.4$, $P < 0.02$
Involuntary at last admission	26	16	$\chi^2 = 5.5$ $P<0.02$
Live alone	26	16	$\chi^2 = 5.5$, $P<0.02$
Single	21	20	NS
Separated/divorced/widowed	11	9	NS
	5	8	NS
Unemployed	34	30	NS
Alcohol and/or drug abuse	16	15	NS
Received antidepressants	16	13	NS
Received lithium	6	6	NS
Received ECT	13	9	NS

TABLE 4

Comparison of schizophrenic suicide victims and schizophrenic controls

Victims	Schizophrenic suicide controls N=28	Schizophrenic Significance N=13	
Past suicide attempt	17	2	$\chi^2 = 7.3$ P < 0.007
Involuntary at last admission	19	6	NS
Live alone	20	6	$\chi^2 = 2.45$ P<0.12
Age	39.0± 14.1	37.2± 10.1	NS
Male	20	9	NS
Female	8	4	NS
Single	20	8	NS
Separated/divorced/widowed	5	3	NS
No psychiatric admissions	6.7± 14.1	7.3± 6.3NS	
Length of last inpatient stay (days)	588.1+ 973.6	170.4 + 196.2	NS
Unemployed	27	11	NS
Alcohol and/or drug abuse	12	3	NS
Received antidepressants	9	5	NS
Received lithium	4	2	NS
Received ECT	11	6	NS

et al. (1972) and Beisser and Blanchette (1961) from the United States, the most common diagnosis was also schizophrenia – 50.5% and 45%, respectively. In England, Goh et al. (1989) reported that schizophrenia was the most common diagnosis being present in 27 of 57 (47.3%) consecutive inpatient or day-patient suicides. Similarly Wolfersdorf and Vogel (1987) in a survey of the 249 inpatient suicides in six state psychiatric hospitals in Germany reported that 132 (53%) were schizophrenic and 62 (25%) depressives. Miles (1977) estimated that approximately 10% of schizophrenics die by committing suicide. However, the precise composition of any series of psychiatric inpatient suicides is likely to vary depending on the way patients are selected for admission to the hospital in question.

It is also noteworthy that 12 of the 37 (32.4%) suicide victims in the present study committed suicide after having been in hospital over one year. This is very similar to the 31. 2% and 26% in the series of Sletten et al. and Beisser and Blanchette, respectively. Others have reported similarly (Crammer, 1984). In the present study, ten of the 12 (83.3%) suicide victims who had been in hospital over a year suffered from schizophrenia. Goh et al. also noted that six of their eight (75%) suicide victims who had been in hospital over a year were schizophrenic. That some schizophrenic patients may commit suicide during the later stages of admission is further suggested by the fact that only two of the 28 schizophrenia suicide victims in the present series committed suicide within the first two months of admission compared with five of the nine suicide victims with other diagnoses (7.1 vs. 55.6%). A proximal factor to the suicide in several of these long stay patients was discharge planning, as part of rehabilitation, which had led to the painful realization by the patient that they were losing the hospital and staff and/or that their family was not prepared to have them home. Morgan and Priest (1984) have drawn attention to "the process of terminal or malignant alienation which may be provoked by recurrent relapse or difficult behavior."

Because three quarters of these inpatient suicide victims suffered from schizophrenia, it was decided to also compare the schizophrenic suicide victims with the schizophrenic controls. In this comparison the only significant difference was that more of the schizophrenic suicide victims than controls had made a previous suicide attempt (60.7 vs. 15.4%, $P < 0.007$). Also significantly more of the schizophrenic suicide victims than controls had made two or more previous suicide attempts (8 of 28, 28.6% vs. 0 of 13, 0%, $P < 0.05$). This issue of previous or recurrent suicidal behavior in some schizophrenic suicide victims is of interest as recent longitudinal studies in both schizophrenic and depressed patients have reported that low cerebrospinal fluid concentrations of the serotonin metabolite 5-hydroxyindoleacetic acid are associated with suicidal behavior during follow up (Cooper et al. 1992; Roy et al. 1989; Roy, 1993).

Three of the 37 (8.1%) suicide victims in the present study were epileptics. All three had temporal lobe epilepsy. It is now well recognized that epileptics, particularly those with temporal lobe foci, are vulnerable to develop affective symptoms and have a raised risk of suicide (Barraclough, 1987).

Brockville Psychiatric Hospital (BPH) is similar in the composition of its services to many State and Provincial hospitals in North America and to the mental hospitals found in the United Kingdom and Europe. BPH is situated close to the St. Lawrence River to its south while a railway line runs close to its northern boundary. These aspects of the hospital's location highlight the issue of availability of method for potential suicide victims as 12 of the 37 (32.4%) suicide victims drowned themselves in the river and seven (18.9%) threw themselves in front of a train. Only a minority of the suicides occurred in the hospital wards – as in other series. Analysis of recent behavior in relationships with others in psychiatric wards may be a fruitful area of study.

Acknowledgement

Thanks are due to Sharon Purvis and JoAnn Bolton in clinical record services of BPH for finding and matching the charts.

References

Barraclough B. The suicide rate of epilepsy. Acta Psychiatrica Scandinavica, 1987; 76: 339–345.

Beisser A, Blanchette J. A study of suicides in a mental hospital. Diseases of the Nervous System, 1961; 22: 365–369.

Cooper S, Kelly C, King D. 5-hydroxyindoleacetic acid in cerebrospinal fluid and prediction of suicidal behavior in schizophrenia. Lancet, 1992; 340: 940–941.

Crammer J. The special characteristics of suicide in hospital in-patients. British Journal of Psychiatry, 1984; 145: 460–476.

Goh S, Salmons P, Whittington R. Hospital suicides: are there preventable factors? Profile of the Psychiatric Hospital Suicide. British Journal of Psychiatry, 154: 247–249.

Miles P. Conditions predisposing to suicide: a review. Journal of Nervous and Mental Disease, 1977; 164: 231–246.

Morgan H, Priest P. Assessment of suicide risk in psychiatric in-patients. British Journal of Psychiatry, 1984; 145: 467–469.

Roy A, DeJong J, Linnoila M. Cerebrospinal fluid monoamine metabolites and suicidal behavior in depressed patients: a five year follow up study. Archives of General Psychiatry, 1989; 46: 609–612.

Roy A. Some determinants of affective symptoms in epileptics. Canadian Journal of Psychiatry, 1979; 24: 554–556.

Roy A. Risk factors for suicide in psychiatric patients. Archives of Geneneral Psychiatry, 1982; 39: 1089–1095.

Roy A. Serotonin, suicide and schizophrenia. Canadian Journal of Psychiatry, 1993; 38: 369.

Sletten I, Brown M, Evenson R, Altman H. Suicide in mental hospital suicides. Diseases of the Nervous System, 1972; 33: 328–334.

Wolfersdorf M, Vogel R. Types of hospital suicide and depression: some selected results of a study on suicides committed during psychiatric hospitalization. Crisis, 1987; 8: 37–48.

6. Twin Research Perspective on Suicide and Suicidal Attempts

NANCY L. SEGAL and ALEC ROY

I. Introduction

The bases of suicide and suicidal behavior have been of considerable interest to psychological and medical professionals. A genetic influence on suicidal behavior bas been demonstrated by a modest, but growing number of twin and adoption studies (Roy, 1983; Juel-Nielsen and Videbach, 1970; Tsuang, 1983; Whender, Kety, Rosenthal, Schulsinger, and Lunde, 1986; Statham, Martin, Dunne, Dinwiddie et al. 1995). Collaboration among investigators at the Twin Studies Center at California State University, Fullerton (Segal), the University of Medicine and Dentistry of New Jersey (Roy), and the Department of Psychiatry at the Catholic University of Rome (Sarchiapone) has produced three new studies as part of an evolving twin research program on suicidal behavior. These studies include a classic twin analysis of completed suicide, a study of suicidal attempts among living twins whose co-twins suicided, and a study of suicidal attempts among twins whose co-twins died due to causes other than suicide. These three studies provide additional evidence of genetic influence on suicidal behavior.

The purpose of this paper is to summarize methods and findings from this new series of twin studies. An overview of twin types and twin study methodology is first provided below.

II. Twin Types and Twin Research Methods

A. *Types of Twins*

Identical or monozygotic (MZ) twins form from the splitting of a single fertilized egg between one and fourteen days after conception. Members of MZ twin pairs share all their genes in common and with rare exception (Dalapicolla, Stomeo, Ferranti, DiLecce, & Purpura (1985), are always of

D. De Leo, A. Schmidtke and R.F.W. Diekstra (eds.), Suicide Prevention, 53–61.

the same sex. Environmental factors, therefore, explain observed behavioral and physical differences between MZ co-twins. Environmental influences include prenatal events, such as fetal anastomoses (which can lead to co-twin differences in size and hemoglobin level), perinatal events, such as adverse circumstances surrounding the birth process, and postnatal events, such as accidents or illnesses affecting only one co-twin. Additional discussion of the nature and origin of MZ co-twin differences is presented in MacGillivray, Campbell, & Thompson (1988) and in Segal (1990).

Fraternal or dizygotic (DZ) twins occur following the fertilization of two separate ova by two spermatozoa. DZ twins, thus, share 50% of their genes, on average, by descent. DZ twin pairs may be same-sex or opposite-sex, events that occur with nearly equal frequency (Bulmer, 1970). Differences between DZ co-twins may be associated with genetic and/or environmental influences.

B. Classification of Twins

Accurate classification of twins as MZ or DZ is a critical initial step in the research process. Misclassification of twin pairs can lead to errors in estimating the relative contributions of genes and environment to the variables of interest. The most accurate method for determining whether twin pairs are MZ or DZ is provided by extensive blood-typing analysis. Samples of blood provided by each co-twin are compared across a series of blood group systems (ABO, Rh, MNSs, K), serum proteins (Gc, Hp, Bf, PLG, Gm, Km 1) and zed cell enzymes (PGMI, AcP, EsD, GLO). It has been shown that when co-twins are compared across the blood group factors listed above, in combination with fingerprint ridgecount, ponderal index and cephalic index, the probability of misdiagnosis is less than 0.001 (Lykken, 1978). On rare occasions, DZ co-twins may scare all measured blood groups, an event explained by having common parents. It is for this reason that similarity across all measured blood groups identifies MZ twins with a high level of certainty, but not with complete certainty. However, co-twin differences in blood groups definitively establishes that a given twin pair is DZ.

Physical resemblance questionnaires can be substituted for blood-typing analysis when issues of expense or convenience arise. This procedure shows approximately 95% agreement with results from blood-typing analyses (Lykken, Bouchard, McGue, and Tellegen, 1990). Several easily administered forms are available for use with adults (Nichols and Bilbro, 1966) and children (Goldsmith, 1991). Additional information on this topic is available in Segal (1984, 1990).

C. Twin Research Designs and Strategies

Sir Francis Galton (1875) first recognized the logic underlying the classic twin method: "It is, that their history affords means of distinguishing between the effects of tendencies received at birth, and of those that were imposed by the circumstances of their after lives; in other words, between the effects of nature and nurture" (p. 391). Even though the embryological bases of twinning had not been established at this time, Galton reasoned that meaningful interpretations and applications of twin data were possible only with reference to the distinction between twin types.

The classic twin method compares the magnitude of resemblance between MZ twins to the magnitude of resemblance between DZ twins, with reference to a behavioral or physical characteristic of interest. Greater resemblance between MZ twins than DZ twins supports a genetic contribution to observed variation in that trait. Trait-relevant environmental influences, or factors that affect the dependent variable in predicted ways, are assumed to affect MZ and DZ twins equally, a concept, known as the *equal environments assumption.* A number of empirical studies support this assumption, given the lack of evidence of meaningful associations between similarity in treatment and behavior (Loehlin and Nichols, 1976; Bouchard, Lykken, McGue, Segal, and Tellegen, 1990). Variants of the classic twin design are described in Segal (1990).

Twin resemblance for continuous traits, such as height, is expressed as an intraclass correlation. The intraclass correlation is a ratio of the variance between twin pairs to the total variation, and can be derived from analysis of variance techniques (Plomin, DeFries, and McClearn, 1990). In contrast, twin resemblance for non-continuous or discrete characteristics (e.g., blood type) is expressed as a concordance rate. Various procedures for calculating this value are summarized in Gottesman & Shields (1982) and in McGue (1992). Twin data may require age- and sex-adjustment prior to analysis, given that twin similarity in these variables may inflate estimates of genetic influence (McGue and Boucbard, 1984). Recent advances in analyzing twin data include multiple regression and biometrical modeling procedures (Loehlin, 1992; Neale and Cardon, 1992).

III. Twin Research Program on Suicidal Behavior

Issues, participants, methods, results and implications from three new twin studies of suicide are summarized below. The overall significance of this work and future research directions are then considered.

A. Study 1: Suicide in Twins (Roy, Segal, Centerwall & Robinette, 1990)

Issue. The genetic transmission of psychiatric disorder is a possible confounding factor in evaluating twin data with reference to the question of an independent genetic factor for suicidal behavior. It was, therefore, of interest to undertake both a classic twin study of suicide and a case history analysis to determine if predisposition for suicide may be genetically transmitted independently of psychiatric disorder.

Participants. Participants included 176 twin pairs in which one, or both, co-twins had suicided. Detailed life history information was available for 11 of the pairs; five of these cases were located by one of us (NLS) through the California State University Twin Loss Study (formerly the Minnesota Twin Loss Study), while the other six cases were made available from colleagues and from previous bereavement studies. Suicide victims included 8 males and 5 females, and age at suicide ranged between 20–54 years. Twins were classified as MZ or DZ on the basis of available blood-typing information, responses to a modified form of the Nichols & Bilbro (1966) physical resemblance questionnaire provided by surviving co-twins, or judgments from family members or referring psychiatrists (Roy et al. 1990).

The remaining 165 cases (53 MZ twin pairs and 112 DZ twin pairs) were provided by the National Academy of Sciences-National Research Council Twin Registry. Twins were classified as MZ or DZ according to an algorithm which considered blood-typing, questionnaire responses, and similarity in selected physical characteristics (Beebe & Simon, 1969). The final sample used in this study included 62 MZ twin pairs and 114 DZ twin pairs.

Methods. Concordance for completed suicide was compared between MZ and DZ twin pairs. These data were also examined in combination with twin data from 3 previous twin studies of suicide (Haberlandt, 1967; Juel-Nielsen & Videbach, 1970; Zair, 1981). In addition, one of us (AR) conducted unstructured psychiatric interviews with surviving co-twins and family members in 11 cases to determine the circumstances surrounding the suicide and to obtain possible evidence of family history of suicidal behavior.

Results. A significantly higher proportion of MZ than DZ twin pairs were concordant for suicide: MZ twin pairs: 7/62 (11.3%); DZ twin pairs: 2/114 (1.8%) [Fischer's Exact Test, $P < 0.01$, two-tailed]. Combining these data with findings from three previous studies also showed significantly greater MZ than DZ twin concordance for completed suicide: MZ twin pairs: 17/129 (13.2%); DZ twin pairs: 2/270 (0.7%) [Fisher's Exact Test, $P < 0.001$, two-tailed]. Psychiatric interviews revealed that 11 of the 13 twin suicide victims had received treatment for psychiatric disorder. In addition, twins in 10 of the 11 pairs had a close relative who had been treated for psychiatric disorder.

Comment. The present findings are consistent with previous studies demonstrating genetic influence on suicidal behavior. Life history information on 11 cases confirmed the well-known association between suicide and psychiatric disorder. The precise nature of the genetic mechanism underlying the expression of suicidal behavior requires further study.

B. Study 2: Attempted Suicide Among Living Co-Twins Suicide Victims (Roy, Segal, & Sarchiapone, 1995)

Issue. A study of attempted suicide among the living co-twins of suicide victims was not available in the literature. In view of research evidence of genetic influence on suicidal behavior, it was anticipated that significantly more MZ than DZ twins whose co-twins had suicided would have attempted suicide themselves.

Participants. Participants included a new series of 35 living twins whose co-twins had suicided. Thirty-two twins were identified through the California Twin Loss Study (NLS)), while three twins were referred by colleagues in the United States (AR) and in Italy (MS). The majority of twins were classified as MZ or DZ by scores on the Nichols & Bilbro (1966) physical resemblance questionnaire, with the exception of two cases in which classification was accomplished by self report from the living twin. The sample included 26 MZ twins and 9 DZ twins, who ranged in age between 22–58 years at the time of participation. Co-twins' age at suicide ranged between 12–53 years. MZ twins included 13 males and 13 females, while DZ twins included 1 male and 8 females; 4 of these females were from opposite-sex twin pairs.

Methods. Information about suicidal attempts was obtained by a semi-structured telephone interview administered to twins whose co-twins suicided. Use of a structured interview was judged to be inappropriate as it might have proven distressing.

Results. Suicidal attempts were reported by 10/26 of the MZ co-twins (38.5%), but by none of the 8 DZ co-twins. The twin group difference in suicidal attempts was significant [Fisher's Exact Test, $P < 0.04$, two-tailed]. Eight of the ten twins who had attempted suicide were female. Twenty-one of the thirty-five suicide victims had received treatment for a psychiatric disorder as had 18 of the 35 living co-twins.

Comment. A genetic influence on suicidal attempts was demonstrated by the higher frequency of such events reported by living MZ than DZ twins whose co-twins had suicided. The higher proportion of females than males among twins who had attempted suicide agrees with previous studies of gender-related differences in suicidal behavior (Holden, 1986; Smith & Crawford, 1986).

C. Study 3: Suicidal Attempts Among Living Twins Whose Co-Twins Death Were Non-Suicides (Segal & Roy, 1995)

Issue. Familial resemblance for suicidal behavior has been associated with varying interpretations: (1) A psychodynamic view would focus on the extreme grief following the loss of a loved one as an explanation for familial resemblance in suicidal behavior. (2) In contrast, a biologically-oriented view would underline the contribution of a genetic predisposition to such behavior. A study of surviving MZ and DZ twins *whose co-twins died due to causes other than suicide* was undertaken to address this issue. In particular, a greater frequency of attempted suicide among bereaved MZ than DZ twins would support the first explanation This is because MZ twins tend to share closer social relationships than DZ twins (Segal, 1988). Alternatively, a low frequency of attempted suicide among this twin group, as well as an absence of MZ-DZ differences, would support the second explanation.

Participants. Bereaved twins were participants in the California Twin Loss Study. The sample included 245 twin survivors: 166 MZ twins (55 males and 111 females) and 79 DZ twins (13 males and 66 females); 41 DZ twins were from opposite-sex pairs. Age at the time of participation ranged between 16–94 years, and age at the time of loss ranged between 15–87 years.

Methods. Twins completed a comprehensive Twin Loss Survey which most received by mail. The survey included a Grief Intensity Scale (adapted from Littlefield & Rushton, 1986), the Grief Experience Inventory (Sanders, Mauger, & Strong, 1985) and a Coping Scale (Littlefield, unpublished). Twins were requested to indicate *True* or *False* to items concerning the occurrence of attempted suicide during the first 1–2 months subsequent to the loss of the co-twin.

Results. Attempted suicide was indicated by 2.4% (6/245) of the twins, three of whom were MZ (1.8%, 3/166) and three of whom were DZ (3.8%, 3/79). As anticipated, the twin group difference in suicidal attempts was not significant (Fisher's Exact Test, $P < 0.39$, two tailed). The six twins who attempted suicide were female and all, but one, had experienced the sudden loss of the co-twin. The frequency of attempted suicide was considerably lower than that reported in our earlier study (28.6%) by twins whose co-twins had suicided (Roy et al. 1995).

Comment. Findings from this analysis support interpretations of twin, family and adoption studies which acknowledge a genetic contribution to familial resemblance for suicide and suicidal behavior. This view is not incompatible with explanations that implicate experiential factors in such behaviors. The results suggest, instead, that the loss of a close relative may be associated with suicidal behavior in a predisposed genotype.

IV. Summary and Implications

The three studies summarized above contribute to a growing literature which demonstrates genetic influence on suicide and suicidal behavior. Appreciation for the value of twin study, methodology for addressing key issues in this area will, hopefully, increase among medical and behavioral science investigators. The need to consider the relatedness between survivor and deceased in future research and counseling concerning suicidal behavior is additionally highlighted by this research program. The nature and availability of specialized support groups and other resources in offsetting adverse consequences following certain forms of loss also warrant further consideration.

The present series of studies is not without limitations. The distribution of MZ and DZ twins among Caucasian populations is 1:2, such that an overrepresentation of MZ twins is evident in the three studies. The majority of twin participants were identified as part of a bereavement study of twins. Twin survivors of suicide who are especially bereaved may be more willing to volunteer for research participation, as compared with twins who are less bereaved, or who have access to effective support systems (Segal & Bouchard, 1993; Segal, Wilson, Bouchard, Gitlin, 1995). Continued identification of end cases involving suicidal behavior in twins, as well as prospectively designed studies, can help to evaluate the effects of these factors on the findings.

V. Future Research Directions

Studies of suicidal behavior using population based twin samples would be desirable. Preliminary collaboration on a study of twin loss with investigators associated with the Australian National Medical Health Research Council Twin Registry will enable further testing of hypotheses and replication of findings using a more representative twin sample. Structural equation modeling will be used to develop models of coping among twin survivors of suicide. These models can be compared with models based on data from twins whose co-twins' deaths were non-suicides.

Acknowledgements

This research was supported, in part, by a Hughes Faculty Research Award and a National Science Foundation Career Advancement Award to Dr. Nancy L. Segal (BNS-8709207); the Minnesota Center for Twin and Adoption

Research, at the University of Minnesota; and the Twin Studies Center, at California State University, Fullerton. Dr. Raymond W. Brandt, Director of the Twinless Twins Support Group Intentional and Ms. Mavis Hall, Director of the Lonesome Dove Support Group in Sydney, Australia are acknowledged for invaluable assistance and support. Karen M. Nelson, B.A., Shelly A. Blozis, M.A., Jennifer Giordano, B.A., Came Ortega, B.A. and Elena Alvarez provided research services.

References

Beebe G, Simon A. Ascertainment of mortality in the U.S. veteran population. American Journal of Epidemiology, 1969; 89: 636–643.

Bouchard TJ Jr, Lykken DT, McGue M, Segal NL, Tellegen A. Sources of human psychological differences, The Minnesota Study of Twins Reared Apart. Science, 1990; 250: 223–228.

Bulmer MG. *The biology of twinning in man.* Oxford University Press, Oxford. Dalapicolla, B., Stomeo, C., Ferranti, G., DiLecce, A., and Purpura, M. (1985). Discordant sex in one of three monozygotic triplets. Journal of Medical Genetics, 1979; 22: 6–11.

Galton F. The history of twins as a criterion of the relative powers of nature and nurture. Journal of the Anthropological Institute, 1875; 5: 391–406.

Goldsmith HH. A zygosity questionnaire for young twins: A research note. Behavior Genetic, 1991; 21: 257–269.

Gottesman II, Shields J. *Schizophrenia: The Epigenetic Puzzle.* Cambridge Universitv Press, Cambridge, 1982.

Haberlandt W. Aportacion a la genetica del suicide. Folia Clinica Int., 1967; 17: 319–322.

Holden C. Youth suicide: New research focuses on a growing social problem. Science, 1986; 233: 839–41.

Juel-Nielsen N, Videbach T. A twin study of suicide. Acta Geneticae Medicae et Gemellologiae, 1970; 19: 307–10.

Littlefield C. Coping Scale (unpublished).

Littlefield CH, Rushton JP. When a child dies: The sociobiology of bereavement. Journal of Personality and Social Psychology, 1986; 51: 797–802.

Loehlin JC. *Genes and environment in personality development. Vol. 2.* Newbury Park: Sage Publications, 1992.

Loehlin J, Nichols R. *Heredity. Environment and Personality. A Study of 850 Sets of Twins.* University of Texas Press, Austin, 1976.

Lykken DT. The diagnosis of zygosity in twins. Behavior Genetics, 1978; 8: 437–463.

Lykken DT, Bouchard TJ Jr, McGue M, Tellegen A. The Minnesota Twin Family Registry, Acta Geneticae Medicae et Gemellologiae, 1990; 39: 35–70.

MacGillivray I, Campbell DM, Thompson B. *Twinning and twins.* Chichester: John Wiley, 1988.

McGue M. When assessing twin concordance, use the probandwise not the pairwise rate. Schizophrenia Bulletin, 1992; 18: 171–176.

McGue M, Bouchard TJ Jr. Adjustment of twin data for the effects of age and sex. Behavior Genetics, 1984; 14: 325–343.

Neale MC, Cardon LR. *Methodology for genetic studies of twins and families.* Kluwer Academic Publishers B.V., Dordrecht, 1992.

Nichols RC, Bilbro WC Jr. The diagnosis of twin zygosity. Acta Genetica et Statistica Medica, 1966; 16: 265–75.

Plomin R, DeFries JC, McClearn GE. *Behavioral Genetics: A primer.* 2nd ed. New York: WH Freeman & Co., 1990.

Roy A. Family history of suicide. Archives of General Psychiatry, 1983; 40: 971–4.
Roy A, Segal NL, Centerwall BS, Robinette CD. Suicide in twins. Archive of General Psychiatry, 1991; 48: 29–32.
Roy A, Segal NL, Sarchiapone M. Attempted suicide among living co-twins of twin suicide victims. American Journal of Psychiatry, 1995; 152: 1075–1076.
Sanders CM, Mauger PA, Strong PN Jr. *A manual for the Grief Experience Inventory.* Charlotte, NC: The Center for the Study of Separation and Loss.
Segal NL. Zygosity diagnosis: Laboratory and investigator's judgment. Acta Geneticae Medicae et Gemellologiae Twin Research, 1984; 33: 515–520.
Segal NL. Cooperation, competition and altruism in human twinships: A sociobiological approach. In KB MacDonald (editor), Sociobiological perspectives on human development, 1988: 168–206. Springer-Verlag: New York.
Segal NL. The importance of twin studies for individual differences research. Journal of Counseling and Development, 1990; 68: 612–622.
Segal NL, Bouchard TJ Jr. Grief intensity following the loss of a twin and other close relatives: Test of kinship-genetic hypotheses. Human Biology, 1992; 65: 87–105.
Segal NL, Roy A. Attempted suicide in monozygotic and dizygotic twins whose co-twins' deaths were non-suicides. Paper presented at the *XVIII International Association for Suicide Prevention* June 4–8, 1995, Venice, Italy.
Segal NL, Wilson SM, Bouchard TJ Jr, Gitlin DG. Comparative grief experiences of bereaved twins and other bereaved relatives. Personality and Individual Differences, 1995; 18: 511–524.
Smith K, Crawford S. Suicidal behavior among "normal" high school students. Suicide and Life-Threatening Behavior, 1986; 16: 313–25.
Statham DJ, Martin NG, Dunne MP, Dinwiddie SH, Slutske WS, Bierut L, Madden PA, Bucholz KK, Heath AC. Genetic factors in suicidal thoughts, intentions, and behaviors. Behavior Genetics (Abstracts), 1995; 2: 289–290.
Tsuang M. Risk of suicide in the relatives of schizophrenics, manics, depressives, and controls. Journal of Clinical Psychiatry, 1983; 44: 396–400.
Wender P, Kety S, Rosenthal D, Schulsinger F, Ortmann J, Lunde I. Psychiatric disorders in the biological and adoptive families of adopted individuals with affective disorders. Archives of General Psychiatry, 1986; 43: 923–29.
Zair K. A suicidal family. British Journal of Psychiatry, 1981; 139: 68–69.

7. The WHO/EURO Multicentre Study on Parasuicide

State of the Art and Future Directions

J.G. SAMPAIO-FARIA

Until the 80s suicide rates in Europe were increasing in almost all countries, particularly in those with high rates. When in 1984 all the Member States of WHO in the European Region adopted a common pan-European health policy, popularly known as "Health Targets for Europe", it was agreed to include the reduction of suicide as one of the 38 targets of the policy (target 12), which states that by the year 2000 there should be a reversal of the rising trends in suicide and attempted suicide.

> *Target 12 – Reducing mental disorders and suicide*
> By the year 2000, there should be a sustained and continuing reduction in the prevalence of mental disorders, an improvement in the quality of life of all people with such disorders, and a reversal of the rising trends in suicide and attempted suicide.

In 1986 a WHO working group on preventive practices in suicide and attempted suicide recommended the need for Member States to promote action for the prevention of suicide in particularly at risk groups. Special attention was recommended for improving our knowledge and expertise on attempted suicide through the collection of internationally comparable data on attempted suicide trends and the study of the factors influencing these trends.

As a consequence it was proposed to WHO to launch a coordinated multinational European study on parasuicide. In 1988 the WHO/EURO Multicentre study on parasuicide initiated its research activities, focused on two major areas:

- The monitoring of attempted suicide
- The identification of high risk for suicide characteristics among attempters which could be used for predicting actual suicide.

Since 1988 until now 16 study areas from 13 countries have collaborated in the Study. As you can see in the Map 1 the study areas are concentrated mainly in northern European countries with traditionally high rates of suicide.

D. De Leo, A. Schmidtke and R.F.W. Diekstra (eds.), Suicide Prevention, 63–66.

Map 1.

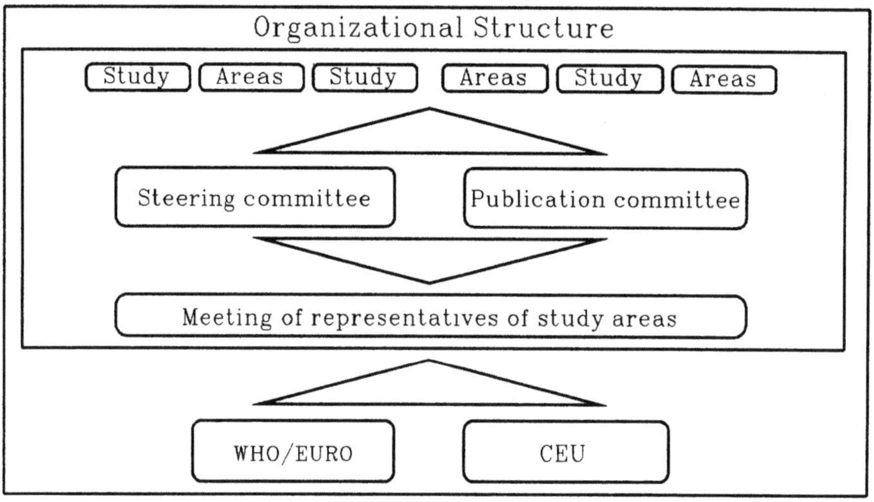

Graph 1.

The research work being developed in these study areas has been supported by excellent and devoted teams of experts and funded by each country's local authorities, research institutions, funding agencies and university departments.

For ensuring its progress, quality of the research activities, and the international coordination of the Study it is organized in the following way (see Graph 1):

The Steering committee members are Dr. U. Bille-Brahe, Coordinator (Denmark), Dr. A. Kerkhof (The Netherlands), Dr. A. Schmidtke (Germany),

D. de Leo (Italy). Dr. S. Platt was the technical coordinator until 1992. The Publications committee is headed by Professor Lönnqvist (Finland).

The WHO contribution has been mainly to promote the Study internationally, to guide and support the international coordination effort of the Steering committee and to assist participating study areas in obtaining funds when so requested. Since 1994 the Commission of the European Union is financially supporting the Study needs for data collection and analysis in the field of prediction of suicide among attempters.

I. Main Outputs of the Study

- Trends in attempted suicide (since 1989)
- Risk profile of suicide attempters (ready in 1996)
- Consolidation of a European Network of experts
- More than 80 publications on attempted suicide
- Fostering action in suicide prevention
  - more countries interested in suicide (mainly CCEE)
  - new suicide prevention centres created
  - national policies/programmes for suicide reduction

II. Future Developments

Expansion of the Study to the Eastern Countries

Since 1992 the European Region of WHO changed considerably both in the number of Member States (from 32 to 50) and in its shape. Now it includes several Member states in Central Asia and the Caucasus. This obliged the European Office of the WHO to reorient its priority work in order to respond to the demands of the countries of central and eastern Europe.

The mean rate of suicide in these countries is considerably higher than in the countries of eastern Europe. On the other side the expertise and resources available in these parts of Europe regarding suicide prevention are very scarce or even nonexisting in some countries. There is thus an urgent need to expand the Study activities to these countries. In collaboration with the Steering Committee of the Study WHO/EURO and CEU are attempting to integrate in the Study new participating study areas from countries in central and eastern Europe. At present official invitations by WHO have been sent to 7 of them and it is expected that they could take active part in the Study from the beginning of 1996 on. This is really a most welcome development for the Study in the future and the WHO will do everything within its range for assisting in the consolidation of this priority need.

More Western European Countries Concerned with Action for Reducing Suicide

In western Europe there is a growing interest in developing national strategies and new services for the prevention of suicide. This has focused attention on the WHO/EURO Multicentre study on parasuicide, and additional study areas are requesting to participate in it, including areas from countries with traditionally low suicide rates, such as Ireland, Greece, and Portugal. In the context of strenghtening the international collaboration among western European countries in the field of suicide prevention it is expected that especially the Member States of the European Union and its Commission will reinforce their support to the Study and to other initiatives attuned to fight this important public health problem in an appropriate manner.

8. Suicide and Suicide Attempt Rates in Europe, 1989–1993

Rates, Changes and Epidemiological Results of the WHO/EURO Multicentre Study on Parasuicide

A. SCHMIDTKE[1], S. FRICKE[2], B. WEINACKER[2], U. BILLE-BRAHE[1], D. DE LEO[1], A. KERKHOF[1], T. BJERKE[3], P. CREPET[3], C. HARING[3], K. HAWTON[3], J. LÖNNQVIST[1], K. MICHEL[3], A. PHILIPPE[3], X. POMMEREAU[3], I. QUEREJETA[3], E. SALANDER-RENBERG[3], B. TEMESVÁRY[3], D. WASSERMAN[3] and J. G. SAMPAIO-FARIA[4]

I. Introduction

In many European countries, suicidal behaviour constitutes a major public and mental health problem and a considerable drain on resources in both primary and secondary health care settings (Schmidtke et al. 1996 ab). Therefore, in implementing the WHO programme, "Health for All by the Year 2000 (HFA 2000)", the European region of WHO identified prevention of suicidal behaviour as a main target. As part of the action in the implementation of target 12 of the WHO European strategy and to develop indicators for this target, comparative data on rates and trends in attempted suicide in European countries were also required. However, due to cross-cultural differences in the medical treatment of suicide attempters and in research methodology, it has proven almost impossible to make valid comparisons of any aspect of suicidal behaviour between different European countries. In the absence of national data, researchers have been forced to rely on local surveys, which vary considerably in terms of their operational definitions of suicide attempts, the representativeness of the samples, the time span covered, the amount of information gathered etc. In addition, local studies have not always been adequate from an epidemiological standpoint.

Therefore, the WHO/EURO Multicentre Project on Parasuicide has a significant role in relation to the action plan of implementing target 12 and in the development of indicators for this target. The background and the organization of the study have previously been described in detail by Schmidtke (1989a),

D. De Leo, A. Schmidtke and R.F.W. Diekstra (eds.), Suicide Prevention, 67–80.

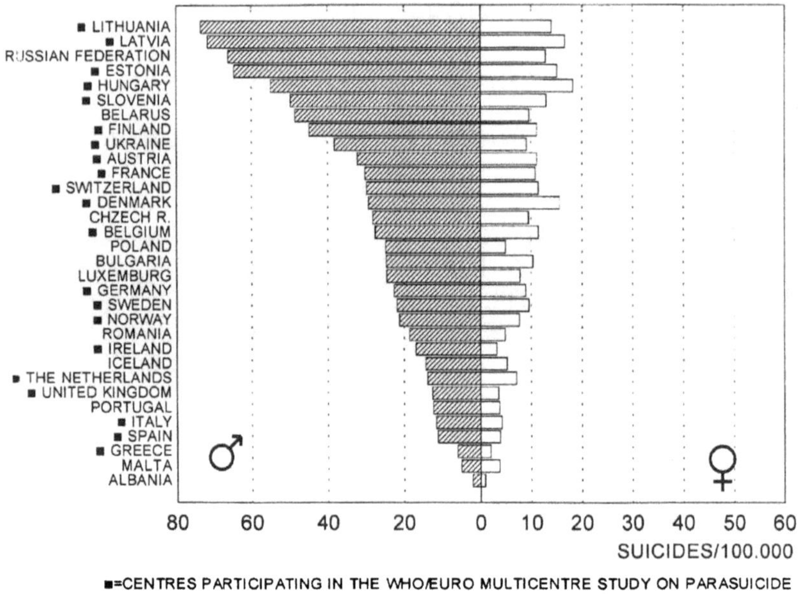

Figure 1. Suicide rates in Europe, 0–99+ years. Sources: WHO, 1994 World Health Statistics Annual (1995).

Platt et al. (1992), Bille-Brahe et al. (1993), Schmidtke et al. (1993) and Kerkhof et al. (1994). A further aim of this project was to generate epidemiological information about attempted suicide in several European countries as well as to gather information about special risk groups. Therefore, in order to avoid the errors made in previous studies as much as possible, a common definition of suicidal behaviour was used. This was the proposed ICD-10 definition. Parasuicide was defined as "An act with nonfatal outcome, in which an individual deliberately initiates a non-habitual behaviour that, without intervention from others, will cause self-harm, or deliberately ingests a substance in excess of the prescribed or generally recognized therapeutic dosage, and which is aimed at realizing changes which the subject desired via the actual or expected physical consequences". The data collection was identical in all centres, using a common monitoring sheet.

II. Suicide Rates

According to the latest official figures released by WHO and the individual National Bureaus of Statistics, the suicide rates among European countries differ widely (Wedel, 1992; WHO, 1995; see Figure 1). However, some regions or states within a country, the former Yugoslavia (Milcinski &

Mrevlje, 1990), Hungary (Moksony, 1995) or Germany (Schmidtke, 1989b; Schmidtke, Weinacker & Fricke, 1996)) for example, often have significantly varying suicide rates.

Since the beginning of the official registration, Hungary has been the country with the highest suicide rates in Europe (if not in the world). The average rate for the period 1989–1993 is 58.8/100,000. Nordic and Eastern (Värnik & Wasserman, 1992) European countries also have somewhat higher suicide rates, while the southern parts of Europe have comparatively low suicide rates. However, in recent years, the rates of some new Eastern countries are surpassing the Hungarian rates (Wasserman, 1996). In all countries, the male suicide rate is higher than the female suicide rate. The average ratio is 3.2 with the lowest male: female ratio of 1.8 in the Netherlands and the highest with 5.1 in Poland (the next highest is found for Belarus with 4.9). Most Nordic and Eastern countries have higher male: female ratios than the Western and Southern countries.

The suicide methods used also differ widely. There are statistically significant preferences for methods of suicide, also in comparison to non-European countries (Snyder, 1994). For example, in Germany "hanging" is the most predominant used suicide method, whereas in some Nordic countries (e.g. Norway), the method "gun" is used in almost a third of all suicide methods. With regard to the drugs and poisons used, there are also great differences in toxicities. For example, the rate of using herbicides and pesticides is much higher in Hungary than in the rest of Europe.

With regard to the age distribution of suicide, one can find a "Hungarian pattern" of the suicide rates in most countries (Wedel, 1992), e.g., the suicide rates increase with age, however, the covariation in males in most countries is more pronounced.

III. Suicide Attempt Rates

Since 1989, 16 centres in 13 European countries have been trying to assess "real" suicide attempt rates and trends as well as the epidemiology of suicide attempts. In this monitoring part of the WHO/EURO Multicentre Study on Parasuicide, until now, these centres are or were participating: Berne, Switzerland; Bordeaux and Cergy-Pontoise, France; Emilia-Romagna and Padova, Italy; Guipuzcoa, Spain; Helsinki, Finland; Innsbruck, Austria; Leiden, The Netherlands; Odense, Denmark; Oxford, U.K.; Sor-Trondelag, Norway; Stockholm and Umea, Sweden; Szeged, Hungary; and Würzburg, Germany.

In 1989, the catchment areas included in total nearly 4,600,000 inhabitants (average population 15+ 1989–1993: 3,970,000)[5]. Figure 2 shows the participating centres.

The registration of suicide attempts was supposed to be carried out in all places where attempters are seen or assessed by health personnel. Because

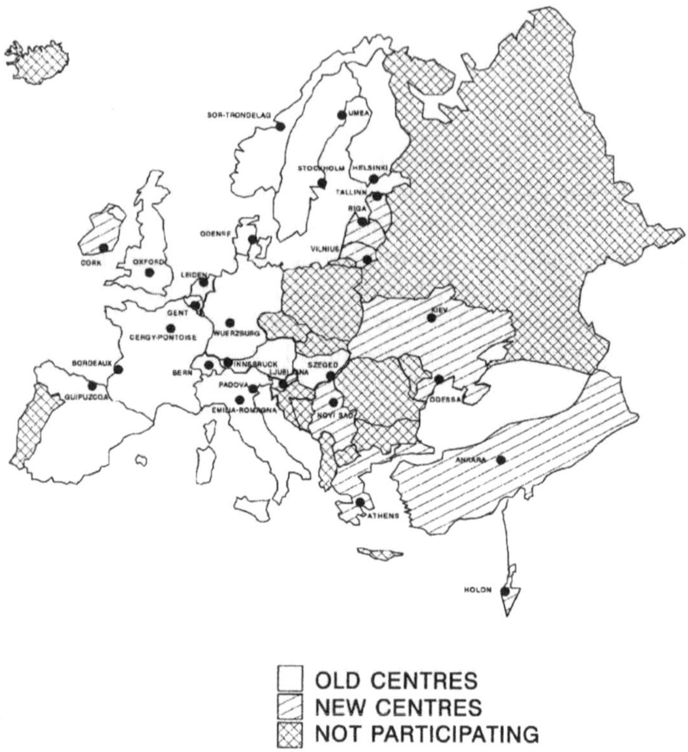

Figure 2. Map of participating centres.

some centres did not cover all local institutions or health agencies the coverage has had to be estimated and confidence intervals calculated. For the computation of the suicide attempt rates, data from all 16 centres were available.

The latest results show the highest average male person rate of suicide attempts (1989–1993) for Helsinki (Finland: 315/100,000), the lowest (46/100,000) for Guipuzcoa (Spain). This represents a sevenfold difference. The highest average female rate was found for Cergy-Pontoise (France: 542/100,000), the lowest (72/100,000) again for Guipuzcoa. With only one exception (Helsinki), the person-based suicide attempt rates were higher among women than men.

In the majority of centres, the highest person-based rates were found in the younger age groups. In half the centres the highest person-based rates were found in 15–24 year-old females (8 of 15 centres; average rate of all centres 282/100,000; the highest average rate in Cergy-Pontoise: 766/100,000, the lowest in Guipuzcoa: 99/100,000). The highest rate for any age group (male or female) in any year was 861 per 100,000 for 15–24 year-old females in

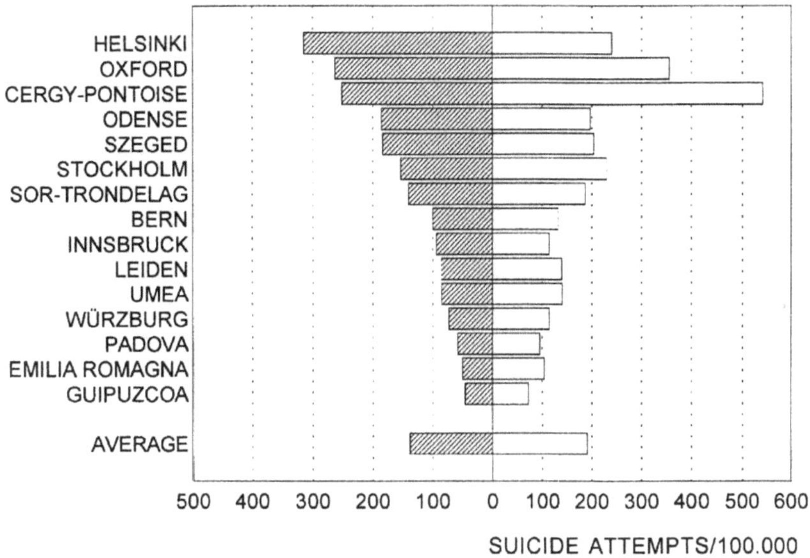

Figure 3. Suicide attempt rates in Europe, 15+ years. Sources: Data of all centres, 1989–1993.

Cergy-Pontoise in 1990. Among males, the highest rates were in age group 25–34 years in 10 of 15 centres. The higest male rate was found in Helsinki with 541 per 100,000 for the age group 35–44 in 1990. In the majority of centres, the rates among people aged 55 years or more were the lowest (males 14 of 15 centres, females 15 of 15 centres).

The highest average person-based ratio of male:female suicide attempt rates was found in the age group 15–24 years (1:1.9), the next highest in the age group 45–54 years (1:1.7). This ratio decreases in the age group 55+ to 1:1.4 – range: 1:3.4 (Cergy-Pontoise) to 1:0.6 (Helsinki).

Even though not all centres provided data for all years where comparisons were possible, the rank order of the different centres did not change drastically over the period covered. Helsinki, Oxford, Cergy-Pontoise, Szeged, and Odense were consistently the centres with the highest rates for males. Emilia Romagna, Guipuzcoa, Padova, and Würzburg were the centres with the lowest rates. The centre with the highest female rates was always Cergy-Pontoise; the second highest was always Oxford, and the third highest rates were most often found for Helsinki. The centre with the lowest rates was always Guipuzcoa followed most often by Padova, Würzburg, and Innsbruck.

For the majority of the centres, the rates for persons aged 15+ years decreased between 1989–1993 (Schmidtke et al. 1996 ab). In 11 centres the male rates decreased between 1989 and 1993 by more than 5%, the average decrease was 14%. The centre with the greatest decrease was Guipuzcoa (69%). However, Innsbruck showed increases (50%) as well as Würzburg

(29%). For the female rates, the average decrease was 10%. 11 centres showed a decrease of more than 5%. The centre with the greatest decrease was Padova (29%). However, the female rates in two centres also showed major increases – Würzburg (75%) and Cergy-Pontoise (7%).

Decreases in rates were also found in most of the individual age groups. The highest average decrease was in the age group 35–44 years for both males (17%) and females (16%). Large decreases were found for females aged 25–34 years (13%) and males aged 15–24 years (15%). The only age group with an average increase was the female age group 55+ (4%).

To examine whether the decreasing trends could be attributed to a "fading out" of the monitoring in the various centres, the trends in the rates for persons aged 15+ years in the individual centres which provided data for five years, were compared. During the period covered, the trends were not uniform. A continuous decrease over the five years was found in only one centre. In most of the centres there were also years with increases. An analysis of the configuration of the trends in the individual centres revealed no unique pattern.

The methods used were primarily "soft" (poisoning) or cutting. 64% percent of the males and 81% of the females used the poisoning method. The second highest method was cutting (mostly wrist cutting). 1.7% of males and 1.2% of females were using solely "alcohol" (X 65) as the method of suicide attempt. Also, the method "car accident" (X 82) is noteworthy (0.5% of the males).

IV. Covariation between Suicide and Suicide Attempt Rates

For the comparison of the suicide attempt rates and the suicide rates, different methods were used. We compared the rank orders and the relations between suicide rates and suicide attempt rates. For three centres, no suicide data for the catchment area were available. For two other centres, due to data protection laws, no suicide data for the individual age groups were available. For all centres the most accessible suicide data for the period 1989–1992 were used.

If the suicide attempt rates in the catchment areas were compared with the suicide rates in the catchment areas (population 15+), the rank orders do not covary significantly for males or for females. For males the mean rank correlation coefficient is 0.45 (n.s.), for females 0.16 (n.s.). The mean ratio suicide:suicide attempt for males is 1:5.3, the range being 1:2.1 (Innsbruck) – 1:19.8 (Oxford; Schmidtke et al. 1995).

The correlations of suicide attempt rates in the catchment area with the suicide rates in the country are higher. The mean rank correlation coefficient for males is 0.51 ($P < 0.05$), for females 0.26 (n.s.). The mean ratio suicide: suicide attempt for females is 1:21.2, the range being 1:7.8 (Szeged) – 1:86 (Oxford; Schmidtke et al. 1995).

V. Sociodemographic Characteristics of Suicide Attempters

For the analysis of the sociodemographic variables during the period 1989–1993, data for 25,343 episodes of attempted suicide from 15 centres in 13 European countries were available. These attempts were made by 19,388 persons. For most of the centres, data for at least the period of 1989–1993 were available. Due to methodological reasons, for each variable, an average of the percentages in each centre was computed.

Only 26% of the males and 32% of the females were married. Nearly half (47%) of all male persons and 40% of all females had never been married.

As the probability of being married and subsequently divorced is lower among younger age groups, a separate analysis for persons older than 24 years was made. This special analysis showed that the general results were age-dependent and that an interaction effect exists. The percentages of single persons diminished with increasing age: only 34% of the males and 22% of the females older than 24 were single. Of those, the percentage who had experienced marital breakdown increased with increasing age: about 20% of the males and females older than 24 years were divorced or separated. Despite this interaction effect, single and divorced people were overrepresented among the suicide attempters.

An average of 36% of the males and 41% of the females were living with a partner or with a partner and children. An average of 28% of the males and 21% of the females were usually living alone. About 5% of the males and females were living in an institution (clinic, sheltered home etc.).

For the international comparison, the different classifications of the national educational categories were recoded into three categories: a low, middle, or high level of education. An average of 57% of the males and females had only a low level of formal education. Only 13% of the males and females had a formal education of the high level. The formal education analysis for the individual centres did not, in general, match the distributions in the general populations according to "educational level", with people of lower educational status being overrepresented among the suicide attempters.

An average of 21% of all male and 12% of all female suicide attempters were unemployed. With regard to economically active persons (those who are able to work), 31% of the males and 24% of the females were unemployed. 34% percent of the males and 48% of the females were economically inactive.

The national categories of social status (level of last/actual job) were recoded in three classes. An average of 52% of the male and 42% of the female suicide attempters in the different centres were in the lowest social class. Only 9% of the males and 5% of the females belonged to the highest category.

Psychiatric diagnoses were collected in four centers. Since the ICD-9 was used at the beginning of the study, the diagnoses were made according to chapter V of this classification system.

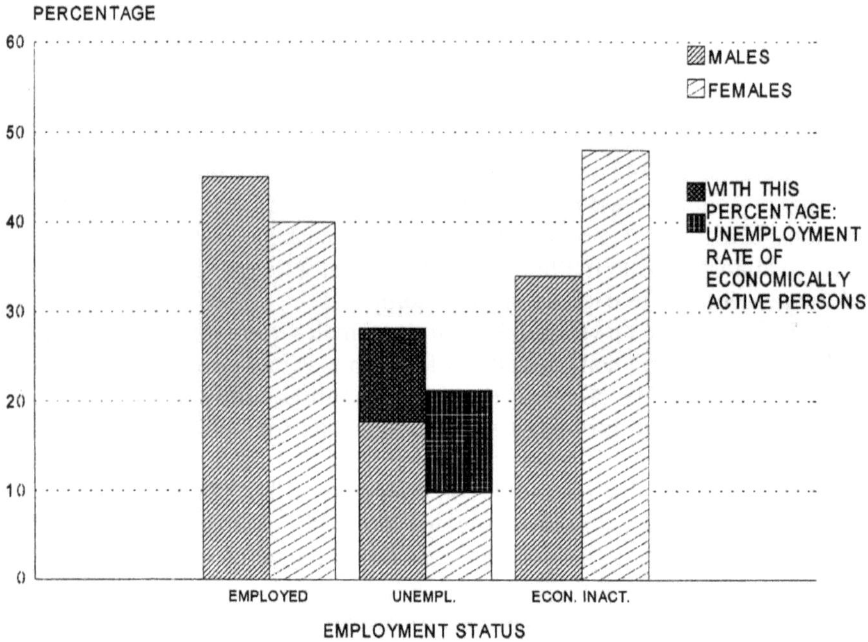

Figure 4. Employment status. Sources: Data of all centres, 1989–1993.

The most prevalent diagnosis for males was "adjustment disorder" (20%; ICD-9: 309). The second highest category was "affective psychosis" (16%). The most prevalent diagnosis for females were "neuroses" (22%; ICD 9: 300). The second highest category for females was "affective psychosis" (21%).

The most often recommended aftercare, about 50%, for both male and female suicide attempters, was "inpatient treatment". For a mean percentage of 16% males and 14% females, no further treatment was recommended after the suicide attempt.

Among males and females, a high percentage of previous suicide attempts can be found. About 40% of both males and females had made a previous suicide attempt. Of the index suicide attempters, about 15% of the subjects committed one or more suicide attempt after they were included in the study.

The analysis of all cases of the total sample, including persons with previous attempts and those with suicide attempts after the monitoring, revealed a general repetition rate of about 50% for both sexes. About 20% of all male and female suicide attempters committed one further suicide attempt within a 12-months period.

Comparisons between repeaters and "first evers" show that the repeaters were more often divorced and more often unemployed. The types of psychiatric diagnoses were also different. While first evers had more "adjustment

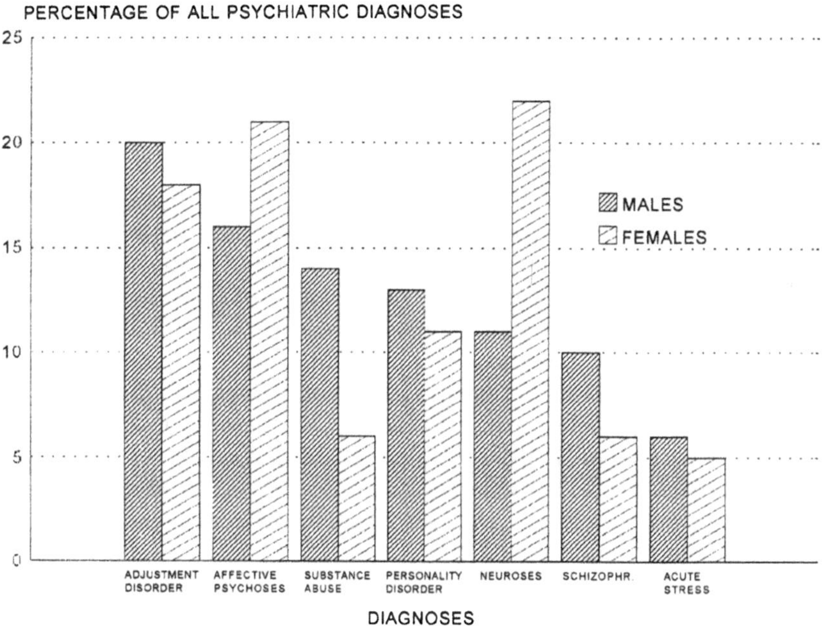

Figure 5. Psychiatric diagnoses. Sources: Data of four centres, 1989–1993.

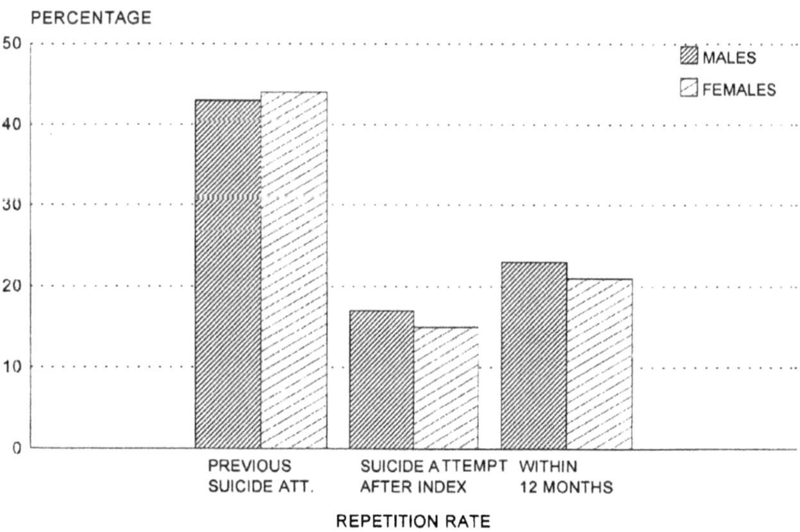

Figure 6. Repetition. Sources: Data of all centres, 1989–1993.

disorders", repeaters had more often received a diagnosis of schizophrenia, addiction or personality disorder.

VI. Discussion

The differences between the various European countries with regard to suicide rates are enormous. Variations in ascertainment procedures and methodological artifacts due to different age structures and percentages of immigrants cannot explain the major part of these variances. However, one of the methodological problems in comparing such international data is that cross-sectional comparisons of national figures on socio-demographic parameters (e.g. unemployment, divorce rates, health facilities etc.) are relatively inadequate to describe and compare people's every day lives and the degree of their alienation and hopelessness, usually seen as factors causing suicidal behaviour (Bille-Brahe et al. 1996). To date, besides some ethnic hypotheses and hypotheses relating to addiction (Wasserman, Värnik & Eklund, 1994), the most convincing hypotheses are those based on different attitudes toward suicidal behaviour (Schmidtke, 1996).

With regard to suicide attempt rates, until now the picture was not so clear. As often noted, cross-sectional and longitudinal analyses of suicide attempt rates in Europe have been hampered by the virtual absence of appropriate national data. Therefore, previous attempts to examine the epidemiology of suicide attempts in Europe were restricted to information gathered in different centres with neither uniform definitions nor standardized case-finding criteria. Consequently, one major objective of the WHO/EURO Multicentre Study on Parasuicide was to gather sufficiently reliable and valid data in order to obtain a credible picture of suicide attempts in the region (both cross-sectional and longitudinal).

The WHO/EURO Study is, of course, not without its methodological problems. However, because of the use of common instruments, definitions and case finding criteria, the WHO/EURO Multicentre Project has for the first time provided the means to compare suicide attempt rates. As a result of the level of expertise and training experience now available at the local level, one can be reasonably confident that the comparisons between centres and the computation of trends which are presented here are reasonably valid. As Platt et al. (1992) stated "... like is being compared with like".

In the majority of the centres it has been estimated that all adult suicide attempts admitted to any health institution in the catchment area have been monitored; and in the other centres, as the various estimation factors show, very few cases appear to have been missed. Due to the near impossibility of establishing a "true" suicide attempt rate in the population, concentration on ensuring the complete coverage of medically treated suicide attempts in the various centres seemed to be the most advantageous method. The various special analyses in individual centres illustrate the validity of this decision.

In previous publications it was already clear that major differences exist between participating European centres. Helsinki, Oxford, Cergy-Pontoise, and Odense lead in the male rates; whereas Cergy-Pontoise, Oxford, Helsinki, and Stockholm lead in the female rates.

The person-based suicide attempt rates for persons aged 15+ years in 7 centres were more than 50% higher among women; in 10 centres more than 30%; in only one centre, Helsinki, was the male rate higher. The average male-female difference was 47%. Similar results could also be found for most of the individual age groups. The greatest differences were found in the younger age groups, whereas the difference diminished with increasing age. However, since the highest male:female ratio for people aged 15+ years was only 1:2.2, the previously mentioned hypothesis that the sex ratio has recently become markedly lower than that found in the seventies seems to be true for a longer period with more reliable data. Further research is required, however, to determine if this is an effect of increasing or stable male rates and/or a general trend for decreasing female rates. So far, the results of the five-year period indicate that the rates are now decreasing in both sexes in the majority of centres, but that in some centres the rates have actually increased (male rates: Innsbruck and Würzburg; female rates: Cergy-Pontoise and Würzburg).

The highest age-specific rates were usually found in the age group 15–24 years for females and in the age groups 25–34 years for males. This confirms the usual view that suicide attempts more commonly occur in teenagers and young adults. However, this simple inverse relationship between age and suicide attempts appears to be less valid for males. Nevertheless, the lowest rates were most often found for the age group 55+ years of both sexes.

The question remains whether these comparisons between the centres are meaningful and valid and whether the results justify statements about similarities and differences in profiles of suicide attempters and trends in suicide attempt rates. However, it does appear unlikely that trends and changes in individual centres could not reflect general trends in the country, although the level of the rates themselves might be different elsewhere.

When decreasing rates have been found in monitoring studies with longitudinal data, it has been suggested that this decrease may be partly due to an effect of a "fading out" of the monitoring (e.g. that the monitoring system gets "fatigued" due to an increasing unwillingness of staff to fill out the monitoring forms). The comparisons of the trends of the rates for persons aged 15+ years, over the five-year period, in the individual centres showed, however, that the trends were not uniform. There was a continuous decrease in only one centre. In most of the centres, there were also years with increases in rates. Therefore, it can be assumed that the decreasing trends cannot be attributed to "fading out" but are reflecting, at least partly, "real" trends. The results indicate that one of the previous goals of the WHO, to reduce the suicide and suicide attempt rates by at least 20 percent, has nearly been reached with regard to attempted suicide. However, it is unclear what factors have led to this decline. Of course, further research efforts are required to establish whether this trend is stable.

The results also show that perhaps some of the long-standing hypotheses and opinions about suicide attempters should be modified. For example, the first results indicate that for some catchment areas the female:male ratio is not as large as found in many previous epidemiological studies on suicidal behaviour.

The methods used for suicide attempts in all centres were primarily "soft" (poisoning) or cutting, as often mentioned in the literature (e.g. Häfner, 1991). The new ICD-10 classification of methods also offered the opportunity to assess the use of some specific methods. Subsequentely, the "alcohol method" seems not to be so rare as previously estimated and the method "car accident" should not be neglected.

With regard to repetition of suicide attempts, the results give the impression that the rate of repetition, in general, is increasing. Among males and females, high percentages of previous suicide attempts can be found with a repetition rate of 20% in one year. This could lead to the hypothesis that suicide attempters are relatively quickly separated into two groups: one group with a single attempt and one group with multiple attempts. These groups differ in a number of variables. Repeaters are more often divorced, unemployed and suffer more often from chronic psychiatric illnesses. They are a high risk group for further suicide attempts and suicides.

This study also provides additional evidence of a detrimental effect of certain sociodemographic or socioeconomic conditions on the occurence of attempted suicide (Fricke et al. 1995; Mösler et al. 1990; Petronis et al. 1990; Rygnestad & Hauge, 1991). Compared to the general population, suicide attempters belong more often to the social categories which indicate social destabilization and poverty. However, economic hypotheses alone cannot explain the differences between the catchment areas (Bille-Brahe et al. 1996). Different values also do not explain the difference in suicidal behaviour (Aish-van Vaerenbergh, 1996). Therefore, for international comparisons, further analyses regarding the base rates of the different variables in the general population, and a cross-cultural framework to account for the data are needed (cf. Leenaars, 1994).

Notes

[1] Steering Group
[2] German WHO Research Group
[3] Principal Investigator
[4] Regional WHO Office, Copenhagen
[5] For all centres the most accessible population data for the period 1989–1993 were used.

References

Aish-van Vaerenbergh A-M. The European Value Study (EVS): theoretical and methodological issues in cross-national or cross-cultural research. Paper presented at the WHO-Meeting "Cross-cultural analysis: Theoretical, methodological and practical issues". Stockholm, 1996.

Bille-Brahe U, Andersen K, Wasserman D, Schmidtke A, Bjerke T, Crepet P, DeLeo D, Haring C, Hawton K, Kerkhof A, Lönnqvist J, Michel K, Philippe A, Querejeta I, Salander-Renberg E, Temesvary B. The WHO-EURO Multicentre Study: Risk of parasuicide and the comparability of the areas under study. Crisis, 1996; 17: 32–42.

Bille-Brahe U, Bjerke T, Crepet P, DeLeo D, Haring C, Hawton K, Lönnqvist J, Michel K, Philippe A, Pommereau X, Querejeta I, Salander-Renberg E, Schmidtke A, Temesváry B, Wasserman D, Sampaio-Faria JG. WHO/EURO Multicentre study on parasuicide. Facts and Figures. Copenhagen: World Health Organization, 1993.

Fricke S, Schmidtke A, Bille-Brahe U, DeLeo D, Kerkhof A, Bjerke T, Crepet P, Haring C, Hawton K, Lönnqvist J, Michel K, Pommereau X, Querejeta I, Phillippe I, Salander-Renberg E, Temesváry B, Wasserman D, Sampaio-Faria J. Sociodemographic characteristics of suicide attempters in Europe. An overview on results of the WHO/EURO Multicentre Study on Parasuicide, 1989–1992. Paper presented at the XVIII International IASP Congress, Venice. (Abstract Book: 33), 1995.

Häfner H. Epidemiologie von Suizid und Suizidversuchen. In: Häfner H (editor), Psychiatrie: Ein Lesebuch für Fortgeschrittene (pp. 210–228). Stuttgart: Gustav Fischer, 1991.

Kerkhof A, Bille-Brahe U, Schmidtke A, DeLeo D, Lönnqvist J (editors). Attempted suicide in Europe. Leiden: DSWO Press, 1994.

Leenaars A. Conference report: Attempted suicide in Europe. Crisis, 1994; 15: 90.

Milcinski L, Mrevlje G. Epidemiologija samoubistva u Jugoslaviji-metodoloska pitanja. Med Pregl., 1990; 43: 453–456.

Mösler TA, Pontzen W, Rummler W. The relationship between suicidal acts and unemployment. In: G. Ferrari, M. Bellini & P. Crepet (editors), Suicidal behaviour and risk factors (pp. 189 - 193). Bologna: Monduzzi, 1990.

Moksony F. The age pattern of suicide in Hungary. Archives of Suicide Research, 1995; 1: 217–222.

Petronis KR, Samuels JF, Moscicki EK, Anthony JC. An epidemiologic investigation of potential risk factors for suicide attempts. Social Psychiatry and Psychiatric Epidemiology, 1990; 25: 193–199.

Platt S, Bille-Brahe U, Kerkhof A, Schmidtke A, Bjerke T, Crepet P, DeLeo D, Haring C, Lönnqvist J, Michel K, Philippe A, Pommereau X, Querejeta I, Salander-Renberg E, Temesváry B, Wasserman D, Sampaio-Faria JG. Parasuicide in Europe: the WHO/Euro multicentre study on parasuicide. I. Introduction and preliminary analysis for 1989. Acta Psychiatrica Scandinavica, 1992; 85: 97–104.

Rygnestad T, Hauge L. Epidemiological, social and psychiatric aspects in self-poisoned patients. A prospective comparative study from Trondheim, Norway between 1978 and 1987. Social Psychiatry and Psychiatric Epidemiology, 1991; 26: 53–62.

Schmidtke A. WHO/EURO Multicentre study on parasuicide. Copenhagen: World Health Organization, ICP PSF 024/18 (WHO paper for the consultation meeting on strategies for reducing suicidal behaviour in the European region). Szeged, 1989a.

Schmidtke A. Suizid- und Suizidversuchsraten in Deutschland. In: Wolfersdorf M, Kaschka W (Hrsg.). Suizidalität – Die biologische Dimension. Heidelberg: Springer, 1989b: 17–32.

Schmidtke A. Suicide: Individual, cultural and international perspectives: Germany/Europe. Paper presented at the 29th Annual Conference of the American Association for Suicidology, St. Louis, Mo, 1996.

Schmidtke A, Bille-Brahe U, DeLeo D, Kerkhof A, Bjerke T, Crepet P, Haring C, Hawton K, Lönnqvist J, Michel K, Philippe A, Pommereau X, Querejeta I, Salander-Renberg E,

Temesvary B, Wasserman D, Weinacker B, Sampaio-Faria J. Suicide and suicide attempt rates in Europe, 1989–1993. Paper presented at the XVIII International IASP Congress, Venice. (Abstract Book: 31), 1995.

Schmidtke A, Bille-Brahe U, DeLeo D, Kerkhof A, Bjerke T, Crepet P, Haring C, Hawton K, Lönnqvist J, Michel K, Philippe A, Pommereau X, Querejeta I, Salander-Renberg E, Temesváry B, Wasserman D, Fricke S, Weinacker B, Sampaio-Faria JG. Attempted suicide in Europe: Rates, Trends and Sociodemographic characteristics of suicide atempters, 1989–1992. Results of the WHO/EURO Multicentre Study on Parasuicide. Acta Psychiatrica Scandinavica, 1996a; 93: 327–338.

Schmidtke A, Bille-Brahe U, Kerkhof A, DeLeo D, Bjerke T, Crepet P, Haring C, Lönnqvist J, Michel K, Salander-Renberg E, Wassermann D, Fricke S, Weinacker B. Results of the WHO/EURO Multicentre Study on Parasuicide: Rates, Trends and Sociodemographic variables of European parasuicides. Paper presented at the WPA Satellite Symposium "Suicidal Behaviour". Budapest, 1996b.

Schmidtke A, Bille-Brahe U, Kerkhof A, DeLeo D, Platt S, Sampaio-Faria J, Henderson J, Pototzky W. The WHO/EURO Multicentre Project on parasuicide – state of the art. Italian Journal of Suicidology, 1993; 3: 83–95.

Schmidtke A, Weinacker B, Fricke S. Epidemiologie von Suizid und Suizidversuch. Nerrenheilkund, 1996; 15: 496–506.

Snyder ML. Methods of suicide used by Irish and Japanese samples: a cross-cultural study from 1964 to 1979. Psychological Reports, 1994; 74: 127–130.

Värnik A, Wasserman D. Suicides in the former Sovjet republics. Acta Psychiatrica Scandinavica, 1992; 84: 229–232.

Wasserman D, Värnik A, Eklund G. Male suicides and alcohol consumption in the former USSR. Acta Psychiatrica Scandinavia, 1994; 89: 306–313.

Wasserman D. Changing suicide rates in the former Soviet Union and in the Baltic countries. Paper presented at the WPA Satellite Symposium "Suicidal Behaviour". Budapest, 1996.

Wedel H. Mortality data in the states around the Baltic. Paper presented at the Symposium "Epidemiology around the Baltic Sea". Stockholm: Royal Medical Society, 1992.

WHO. 1994 World Health Statistics Annual. WHO, Genève, 1995.

9. Relevance of Diagnostic Setting in Predicting the Outcome of Suicide Attempters

PAOLO SCOCCO and DIEGO DE LEO

I. Introduction

Psychiatric disorders, both past and present, appear to be a very significant variable in influencing completed and attempted suicidal acts. According to numerous studies, they are present in over 90% of suicides [1, 2, 3] and between 70% and 90% of all parasuicides [4, 5]. The possibility of predicting suicide in a subject for whom a psychiatric disorder has been diagnosed, however, varies according to the context in which the diagnosis has been made [6, 7]. The aim of this work, based on a group of parasuicidal subjects monitored during the period from January 1989 through December 1992, was to assess the characteristics of individuals with or without a psychiatric diagnosis (groups 1 and 2, respectively) in relation to subsequent evolution. Amongst the patients with a psychiatric diagnosis we distinguished: i) the characteristics of those not diagnosed as suffering from a psychiatric disorder during psychiatric consultation, but diagnosed as having a psychopathology on admittance to a psychiatric facility or during outpatient contact/treatment (subgroup 1A); ii) the characteristics of patients who were diagnosed a psychiatric disorder for the first time during consultation, not having any other form of contact with psychiatric services (subgroup 1B); iii) the characteristics of patients who had been diagnosed a psychiatric disorder during both consultation and an inpatient or outpatient contact/treatment regimen (subgroup 1C).

II. Materials and Methods

The study was performed by the Suicidology Unit of the Department of Neurology and Psychiatry of the University of Padua. The subjects considered in this research – aged fifteen and over – have already been studied in a previous work [8, 9]. The above Unit provides a suicidological consultation service

D. De Leo, A. Schmidtke and R.F.W. Diekstra (eds.), Suicide Prevention, 81–91.

at the local general hospital, in addition to identifying cases of parasuicide in the city of Padua, as part of the Monitoring Phase of the international "WHO/Euro Multicentre Study on Parasuicide" [10]. During suicidological consultation at the local general hospital, a psychiatric diagnosis may be formulated according to DSM-III-R [11] criteria. A psychiatric diagnosis was formulated in 49% of all patients considered (233 out of 474), but only 60% of all cases were referred for suicidological consultation (n=282). Hence, a psychiatric pathology was diagnosed for 83% of parasuicidal patients taking part in suicidological consultation. Various reasons prevented suicidological consultation in the remaining 192 subjects, namely: a) discharge (or self-discharge) immediately after admittance (n=93); b) discharge after first aid treatment at the Accident and Casualty Department or Emergency Psychiatry Department (n=59); c) immediate admittance, following attempted suicide, to a psychiatric ward (n=35); d) the parasuicide was reported by an outside source (GPs) (n=5).

A retrospective survey was performed on the basis of case history data available from 1984, in addition to a follow-up study through to April 1994 (mean follow-up: 40 months, range 16–64 months). Inpatient data were collected from the files of the Local Health Service of Padua and the only private psychiatric clinic operating in the Padua district, while outpatient data were taken from the files of the 9 community-based psychiatric units of the Padua Local Health Service. The study did not include private psychiatrists (which may, to a minor extent, have influenced the number of subjects in subgroup 1B).

Discharge diagnoses provided by the psychiatric services were formulated on the basis of ICD9 criteria [12]. For the purposes of this research, these were converted into the homologous DSM-III-R diagnoses [11]. The psychiatric disorders present in the sample have been divided up into seven diagnostic groups (see description below). Statistical analyses were performed using Student's t, chi-square, Fisher, and Anova tests from the SYSTAT statistical package [13], where appropriate.

III. Results

In addition to the diagnoses formulated during consultation in the period 1989–1992, on retrospective and prospective examination it emerged that a psychiatric diagnosis had been formulated at some point in the test period (1984–1994) for 74% of the 474 (n=351) subjects considered.

If we then consider only the subjects with a psychiatric diagnosis (group 1) and cluster them as described in the introductory part of the work, subgroup 1A includes 118 subjects (with a psychiatric diagnosis formulated in a setting other than consultation); subgroup 1B (psychiatric diagnosis formulated during consultation only) consists of 90 patients; and subgroup 1C (psychiatric diagnosis formulated in more than one setting) is composed of 143 individuals.

In the overall sample there was a prevalence of females (67%, n=316) over males (33%, n=158); the difference in distribution between groups 1 and 2 (64% vs 36% in group 1; 75% vs 25% in group 2, respectively) was statistically significant (chi-square = 4.941, $P < 0.03$).

The mean age for the patients in group 1 (n=351) was 40.48 (S.D. 17.60) and 36.37 (S.D. 18.49) for those in group 2 (n=123), (t=2.145, $P < 0.03$). On analysis of the sex and age variables among the subgroups with psychiatric diagnosis (subgroups 1A, 1B and 1C), no statistically significant differences were found. As for occupation at the time of parasuicide, broken down into low, middle, high occupational levels and unemployed (in line with the protocol of the WHO/EURO multicentre study on Parasuicide; [10], a statistically significant difference emerged between groups 1 and 2 (chi-square=10.117, $P < 0.01$). The subjects in group 1 more frequently had a low occupational level and the majority of group 2 patients were unemployed. No statistically significant difference emerged for this variable in subjects in the psychiatric diagnosis subgroups and there were no significant differences in relation to household composition. Indeed, in the groups and subgroups the majority of subjects lived with their parents or with a partner and children. The same applies to marital status, with no significant differences. Educational level was middle/low in subjects with a psychiatric diagnosis, and middle/high in group 2 subjects, with a tendency towards statistical significance (chi square=8.931, $P < 0.07$). There were no statistically significant differences between the subgroups in relation to this variable. Assessment of the methods used in parasuicides, divided up into "soft" methods (drug poisoning and superficial wrist cutting) and hard methods, showed that hard methods were significantly more frequent in group 1 than in group 2 (10.9% vs 4.5%, n=35 vs 5) (chi-square=4.020, P = < 0.05). Significant differences did not emerge when the same analysis was performed in the three subgroups of subjects with a psychiatric diagnosis. 60 of the 474 subjects (12.6%) exhibited repeated suicidal behaviour, of whom 58 belonged to group 1 (chi-square=18.28, $P < 0.000$). The same variable (repetition) also showed significantly different distribution in the case of the subgroups with a diagnosis: 14 in group 1A (12%), 4 in 1B (4%), and 40 in 1C (28%) (F=52.001, p=<.000).

Distribution of the various diagnoses of the 351 subjects is given in Table 2. Mood disorders prevailed in all subgroups (1A, 1B, 1C).

While subgroup 1C consisted of 143 subjects, Table 2 reports data based on only 78 patients. This is because the psychiatric diagnoses for the remaining 65 subjects were conflicting and discordant. As different from other settings, which tend to be characterized by a more enduring relationship, during consultation, mood disorders were diagnosed more frequently than other disorders (see Table 3).

TABLE 1

Characteristics of parasuicidal subjects with or without psychiatric diagnosis

	Group 1 Subjects with diagnosis (n=351)	Group 2 Subjects without diagnosis (n=123)	P
Mean age	40.48 (DS=17.60)	36.37 (DS=18.49)	<0.03
F vs M (%)	64 vs 36	75 vs 25	<0.02
Hard method	10.9%	4.5%	<0.05
Repeaters	16.5%	1.6%	<0.000
Fam. history psych. pathol.	38%	18%	<0.005
Preval. schooling level	Middle/low	Middle/high	n.s.
Occupational level	Low (– unempl.)	Middle/high (+ unempl.)	<0.01
Household composition	With parents/ partner+child	With parents/ partner+child	n.s.
Marital status	single/married	single/married	n.s.

TABLE 2

Distribution of diagnostic groups

Diagnostic group	Subgroup 1A Diagn. after inp/outp trtmt	Subgroup 1B Diagn. after consultation	Subgroup 1C Diagn. after cons-inp/outp trtmt*
Schiz. and other psych. disorders	19% (n=23)	2% (n=2)	28% (n=22)
Mood disorders	**36% (n=42)**	**50% (n=45)**	**47% (n=37)**
Anxiety disorders	6% (n=7)	3.5% (n=3)	0% (n=0)
Subst. abuse/depend. disorders	6% (n=7)	12% (n=11)	3% (n=2)
Adjustment disorders	8% (n=9)	12% (n=11)	3% (n=2)
Other diagnoses	6% (n=7)	3.5% (n=3)	3% (n=2)
Personality disorders	19% (n=23)	17% (n=15)	16% (n=13)
Total	100% (n=118)	100% (n=90)	100% (n=78)

* Subjects with discordant diagnoses after consultation and inpatient/outpatient treatment were not considered

Since intervention takes place relatively early (generally a few hours after admittance to hospital) and, more important, in the absence of an opera-

TABLE 3

Diagnostic discrepancies in subjects in subgroup 1C

Diagnoses after consultation	Diagnoses after inp/outpatient trtmt	% (n)
Mood disorders	Schiz. and other psych. disord.	15% (n=10)
Mood disorders	Personality disorders	11% (n=7)
Mood disorders	Other diagnoses	11% (n=7)
Mood disorders	Anxiety disorders	8% (n=5)
Personality disorder	Mood disorders	11% (n=7)
Other diagnoses	Other diagnoses	44% (n=29)
		100% (n=65)

tional computerized network, diagnosis is rarely influenced by information on patients' case histories. For this reason (chiefly due to endemic disorganization) the diagnostic process during consultation is to all extents and purposes isolated and hence methodologically suitable for comparison with other traditional diagnostic settings, which otherwise inevitably tend to mutually contaminate each other. The exception to this rule were 11 patients who informed staff that they were in the care of the community-based psychiatric services and hence their clinical record cards could be checked during consultation. Said subjects are listed in subgroup 1C and are not included in the group of diagnostic discrepancies. Moreover, on the subject of diagnostic discrepancy, it should be pointed out that 26 of the patients in subgroup 1A (subjects with a diagnosis formulated during inpatient/outpatient care only) were referred to our unit, but no structural diagnosis was made for them.

IV. Other Variables

Family history of psychiatric disorders was present in 38% (n=88) of subjects with a diagnosis, as opposed to 18% (n=10) without a diagnosis (missing=187, 39%; chi-square=7.712, $P < 0.005$). In subgroups 1A, 1B and 1C, however, there were no statistically significant differences for this variable. Family history of suicidal behaviour was not, instead, statistically different between the groups (or subgroups) considered.

At the time of the parasuicide, 19% of the whole sample also exhibited a somatic pathology (e.g. diabetes, polyarthrosis, etc.) (missing=166, 35%), with uniform distribution between groups 1 and 2. If group 1 is broken down into subgroups 1A, 1B and 1C, there is a somatic pathology present in 33%

TABLE 4

Parasuicidal subjects by therapy adopted prior to parasuicide

	Group 1	Group 2
	Subjects with diagnosis	Subjects without diagnosis
No therapy	23% (n=54)	**45% (n=24)**
Benzodiazepines	**28% (n=68)**	23% (n=12)
Antidepressants	18% (n=42)	15% (n=8)
Neuroleptics	21% (n=50)	2% (n=1)
Other therapies	10% (n=24)	15% (n=8)
Total	100% (n=238)	100% (n=53)

Missing=183

* chi-square=19.007 DF=4 P < 0.001

(n=16) of group 1A subjects, 17% (n=21) from group 1C and 13% (n=10) from group 1B (chi-square 8.309 P < 0.02).

As concerns the patients' last visit to his/her G.P. prior to the parasuicide, there were no statistically significant differences between the various groups: roughly 1/3 of subjects had seen the G.P. during the week leading up to the event, and a further 1/3 in the month prior to the event (missing 214, 51%). The drug therapy undertaken by patients during the period prior to the parasuicidal act varied significantly for groups 1 and 2 (Table 4), but not for subgroups 1A, 1B and 1C.

As for the pharmacological treatment adopted during the period prior to the attempted suicide, combined therapies are not taken into consideration, rather a hierarchical criterion (NL>AD>BZ>other therapies) is utilized. These treatments had, in the majority of cases, been prescribed by psychiatrists for group 1 subjects, particularly those from subgroups 1A and 1C, and by G.P.s for group 2 subjects (significantly different from group 1; chi-square 19.007, P < 0.001).

On the basis of the mean follow-up (40 months), the percentage of completed suicides in the total sample was 2.3% (11 cases). 8 of these 11 cases belonged to group 1 (2.3%) and 3 to group 2 (2.4%). On analysis of the suicides in group 1, moreover, it emerged that 6 were part of subgroup 1C (4.2%) and 2 of subgroup 1A (1.7%). There were no suicides in subgroup 1B.

V. Discussion

During the time span considered by the study, the percentage of subjects with a psychiatric diagnosis was decidedly higher than data provided by the Padua centre involved in the WHO/EURO Multicentre Study on Parasuicide (74% vs 49%)[9], although it was lower than the percentage emerging from consultation data only (83%). The discrepancy between the first two percentages may be accounted for by the need, in the WHO study, to check and report all cases of parasuicide occurring within the assigned catchment area, irrespective of the source and completeness of the data available. The most problematic aspect of this work clearly concerns the percentage of psychiatric diagnoses formulated during consultation. The number appears to exceed the total percentage of diagnoses resulting from the overall sample considered by this research. Various considerations must be made on this subject. First and foremost is the discrepancy emerging between diagnoses made during consultation and those formulated in other diagnostic settings (inpatient/outpatient treatment) during which time there may clearly be several opportunities to verify the diagnostic process initiated.

At first sight, this observation may suggest that many diagnoses formulated during consultation are incorrect and, besides highlighting the limits of the diagnostic process in general and "isolated" diagnoses in particular [14], it may imply the following: subjects who have just attempted suicide tend to present a depressive clinical picture. This may prevent the consultant psychiatrist from noting the presence of other underlying psychopathologies; the psychopathological picture may evolve (quantitatively and/or qualitatively) during the personal history of a given patient [14]; individuals who carry out acts of self-injury may present comorbid pathologies [1, 15, 16]. This discrepancy in figures can, however, also be justified by the fact that the specificity of the consultation setting with suicidal subjects brings into play a dishomogeneous array of factors, such as: closeness to the peak of the crisis precipitating the parasuicidal act; a reworking through (apparently often depressively) of the act; the alien nature of the hospital environment; contact with a doctor (psychiatrist) unknown to the subject and often considered disagreeable [17]; the place where consultation takes place is frequently inadequate (e.g. in close proximity to other patients etc.); psychiatrists' habit of classifying suicidal acts as typical of depressive disorders (suicidal ideation is among the diagnostic criteria for classifying major and "minor" depression). For these reasons, and probably others, diagnosis of the clinical picture which emerges in this setting is, as we have ascertained in our study, subsequently disconfirmed in several cases.

Table 3 shows that in 29 cases, a diagnosis of a mood disorder made during consultation was modified in other health care settings, while in only 7 cases diagnosis of personality disorder was, rather surprisingly, changed to a mood disorder completely different from the assessment made during consultation.

Moreover, in 26 cases, a structured clinical diagnosis was not formulated in consultation, despite the fact that the subjects belonged to subgroup 1A. The majority of said subjects had personality disorders and this anomaly can be accounted for by the fact that during consultation, the diagnostic approach to patients is oriented around psychoclinical examination, which is not traditionally targeted at making Axis II diagnoses.

Emphasis should be laid on the fact that 28% of subgroup 1C have schizophrenic disorders while this is the case in only 2% of subgroup 1B members; this accounts for the lack of psychiatric history among subjects in the latter group.

In the surveyed sample, some cases of schizophrenic disorder were mistaken during consultation for a mood disorder. In such cases, it is clearly possible to develop a therapy which does not permit correct management of the patient and potentially influences the risk of the act being repeated or the quality of life being adversely affected by reason of the incorrect therapy prescribed. It should nonetheless be borne in mind that suicidal crises in schizophrenics often arise during periods of depression [18], which may account for the type of diagnosis made during consultation.

From this research it may be deduced that improper or unconfirmed diagnosis/treatment could be avoided by providing adequately equipped, more widely-embracing inpatient care, where patients are not simply admitted to hospital (generally with a negative attitude or without receiving particular attention from medical staff) and examined by unsuitably trained personnel, often resulting in patients being discharged with vague therapeutic directions or even none at all [19].

There was a relatively low number of comorbid somatic pathologies in the study sample compared to other studies [20]. However, in subjects from subgroup 1A, the fact that a high frequency of somatic pathologies were identified (33%), may have masked comorbid psychiatric symptomatology, partly explaining the lack of a diagnosis during consultation or failure to refer subjects in this subgroup, who were considered to be suffering mainly from somatic pathologies.

The lower schooling and occupational level in subjects in group 1 may be due to the impact that psychiatric pathologies have on individual performance. The higher frequency in the use of hard parasuicidal methods and repeated suicidal behaviour among members of group 1 denote their "greater severity". From our study it emerged that family history of psychiatric pathology rather than of suicidal behaviour appears to distinguish parasuicidal subjects with a psychiatric pathology from those with no such pathology. Subjects receiving a diagnosis form a rather homogeneous group and the only characteristics which distinguish them are reported in Table 5.

The mean follow-up (40 months) showed a percentage of completed suicides of 2.3% of the total sample (11 cases), in line with data by Hawton (21) and Suokas & Lonnqvist [22], but lower than the figures reported in

TABLE 5

Distinguishing characteristics between parasuicidal subjects with diagnosis (subgroups 1A-B-C)

	Subgroup 1A Diag. after inp/outp trtmt	Subgroup 1B Diag. after consultation	Subgroup 1C Diag. after cons- inp/outp trtmt	P
Suicides	1.7%	0%	4.2%	n.s.
Repeaters	12%	4%	28%	<0.000
Concomitant somatic pathol.	33%	13%	17%	<0.01

other works [23]. As previously reported, of the 11 suicides, 8 belonged to group 1, and 3 to group 2. The absence of suicides in subgroup 1B corroborate the findings of Allebeck & Allgulander [6, 7] relating to the limits of "isolated" diagnoses in predicting a future completed suicidal act. Hence, psychiatric pathology appears to be an important factor, in both parasuicide and completed suicide, where this has been diagnosed longitudinally [6, 7, 24].

VI. Conclusions

This study confirms the importance of psychiatric pathology in general, and depressive pathologies in particular, in the context of parasuicidal behaviour. The percentage of parasuicidal subjects who have completed suicide falls within the range of values reported in previous studies, despite the fact that the total number of suicides in our sample proved to be relatively low. Although no differences between suicides performed by subjects with or without psychiatric diagnosis emerged in our study, importance should be given to the "quality" of the diagnosis itself. One of the limits of this study may be the lack of comparison between diagnoses formulated in different settings but close in time to the parasuicidal act. Nevertheless, the fact that there were no completed suicides among subjects who only received a diagnosis during consultation suggests on the one hand that the importance of psychiatric pathology cannot be assessed with sufficient clarity in this particular context, and on the other hand, may cast doubts on the utility of studies based exclusively on diagnoses formulated in this particular setting for predicting subsequent suicidal behaviour.

References

1. Henriksson MM, Aro HM, Marttunem MJ et al. Mental disorders and comorbidity in suicide. American Journal of Psychiatry, 1993; 150: 935–940.
2. World Health Organization, Division of Mental Health. Guidelines for the primary prevention of mental, neurological and psychosocial disorders. WHO/MNH/MND/93.25, Geneva, 1993.
3. Lesage AD, Boyer R, Grunberg F et al. Suicide and mental disorders: A case-control study of young men. American Journal of Psychiatry, 1994; 151: 1063–1068.
4. Morgan HG, Burns-Cox CJ, Pocock H et al. Deliberate selfharm: Clinical and socioeconomic characteristics of 368 patients. British Journal of Psychiatry, 1975; 127: 564–574.
5. Urwin P, Gibbons JL. Psychiatric diagnosis in self-poisoning patients. Psychological Medicine, 1979; 9: 501–507.
6. Allebeck P, Allgulander C. Psychiatric diagnoses as predictors of suicide. A comparison of diagnoses at conscription and in psychiatric care in a cohort of 50,465 young men. British Journal of Psychiatry, 1990a; 157: 339–344.
7. Allebeck P, Allgulander C. Suicide among young men: Psychiatric illness, deviant behavior and substance abuse. Acta Psychiatrica Scandinavica, 1990b; 81: 565–570.
8. De Leo, D, Fantinato S. The epidemiology of attempted suicide in Padua, 1989–1991. AJFM Kerkhof, A Schmidtke, U Bille-Brahe et al (editors): Attempted Suicide in Europe. DSWO Press, Leiden University, 1994.
9. Fantinato S, Bernardini M, Padoani W et al. Epidemiologia dei parasuicidi a Padova, 1989–1992. D De Leo, W Padoani & S Fantinato (editors): Comportamenti Suicidari a Padova, 1989-1992. Kendall, Padova, 1994.
10. Platt S, Bille-Brahe U, Kerkhof A et al. Parasuicide in Europe: the WHO/EURO multicentre study on parasuicide. I; Introduction and preliminary analysis for 1989. Acta Psychiatrica Scandinavica, 1992; 85: 97–104.
11. Diagnostic and Statistical Manual of Mental Disorders, 3rd ed, revised. American Psychiatric Association, Washington, DC, American Psychiatric Press, 1987.
12. World Health Organization. Manual of the International Statistical Classification of Diseases, Injuries, and Causes of Death, 9th revision. Vol 2, (Alphabetical Index). WHO, Geneva, 1977.
13. Wilkinson L. SYSTAT, Inc. Evanston, IL, 1989.
14. Fennig S, Kovasznay B, Rich C et al. Six-month stability of psychiatric diagnosis in first-admission patients with psychosis. American Journal of Psychiatry, 1994; 151: 1200–1208.
15. Rudd MD, Dahm PF, Rajab MH. Diagnostic comorbidity in persons with suicidal ideation and behavior. American Journal of Psychiatry, 1993; 150: 928–934.
16. Angst J. Comorbidity of anxiety, phobia, compulsion and depression. International Clinical Psychopharmacology, 1993; suppl.1: 21–25.
17. Grootenhuis MA, Paap RPSC. How do suicide attempters evalute their treatment in a general hospital. Italian Journal of Suicidology, 1993; 3: 25–28.
18. Roy A. Suicide in schizophrenia. In: Suicide. A Roy (editor) Williams and Wilkins, Baltimora, 1986.
19. Pommereau X. Conduzione dei pazienti tentatori di suicidio nell'ospedale generale. Verso l'implementazione di specifiche regole routinarie di trattamento e di pervenzione a livello europeo. In Suicidio: Verso Nuove Strategie Preventive. D De Leo & L Pavan (editors), Kendal, Padova, 1994.
20. Stenager EN, Stenager E, Jensen K. Attempted suicide, depression and physical diseases: A 1-year follow-up study. Psychotherapy & Psychosomatics, 1994; 61: 65–73.
21. Hawton K. Il completamento del suicidio successivo al tentativo di suicidio. D De Leo (editor): Aspetti Clinici del Comportamento Suicidario. Liviana, Padova, 1990.

22. Suokas J, Lonnqvist J. Outcome of attempted suicide and psychiatric consultation: risk factors and suicide mortality during a five-years follow-up. Acta Psychiatrica Scandinavica, 1991; 84: 545–549.
23. Buglass D, Mc Culloch JW. Further suicidal behaviour: the development and validation of predictive scale. British Journal of Psychiatry, 1970; 116: 483–491.
24. Merikangas KR, Wicki W, Angst J. Heterogeneity of depression. British Journal Psychiatry,1994; 164: 342–348.

10. Gender Differences in Adolescent Suicide*

MAURI J. MARTTUNEN, MARKUS M. HENRIKSSON, HILLEVI M. ARO, MARTTI E. HEIKKINEN, ERKKI T. ISOMETSÄ and JOUKO K. LÖNNQVIST

Previous psychological autopsy studies have reported on correlates of completed suicide among adolescents (Brent et al. 1988; Gould et al. 1992; Marttunen et al. 1992; Shafii et al. 1985), adolescents and young adults (Rich et al. 1986; Runeson, 1989), and specifically among young males (Apter et al. 1993; Lesage et al. 1994). Although females comprise about one fifth of all young suicides in Western countries (Barraclough, 1988), research on youthful female suicides has been scarce. Finland has one of the highest suicide rates both among young males and females (WHO, 1992a; 1992b; 1993). There is some variation between countries in the male to female ratio in completed suicide among youth, but of all suicides in Western countries in the age group 15 to 24 years about 15% to 20% are committed by females (Figure 1).

It has been suggested (Shaffer, 1988) that depression and transient situational problems would be commonly associated with suicide among adolescent females, but that antisocial and aggressive behavior, often with comorbid substance abuse would be more common in young male suicides. In a previous study on suicides in 13 to 19-year old adolescents we reported our clinical impression that the female victims seemed to be more severely mentally disturbed than the males (Marttunen et al. 1991).

The aim of this study was to characterize 13- to 22-year-old female suicide victims in a nationwide unselected youth suicide population and to compare them with male victims in terms of family background, suicide methods, frequency and severity of mental disorders and antisocial behaviors, and history of psychiatric treatment. We expected that the female suicides would be more likely to have previous suicide attempts, major depression and a history of psychiatric care but less likely to have antisocial and substance use disorders and adjustment disorders than the male victims.

D. De Leo, A. Schmidtke and R.F.W. Diekstra (eds.), Suicide Prevention, 93–103.

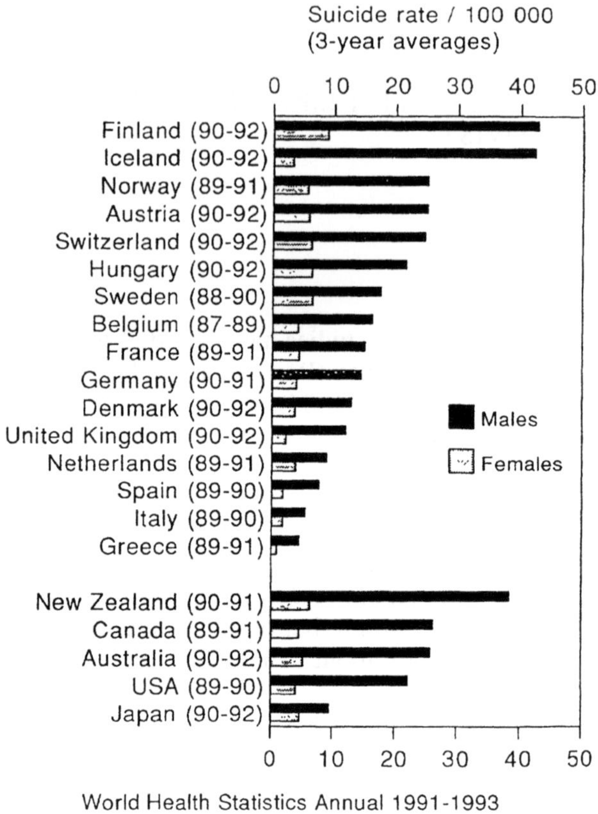

Figure 1. Male and female suicide rates among 15–24 year olds in selected countries

I. Methods

The data were collected during the research phase of the National Suicide Prevention Project in Finland. In the project, all (N=1397) suicides committed in Finland during a 12 month period (April 1st, 1987 and March 31st, 1988) were carefully analyzed using the psychological autopsy method (Lönnqvist, 1988).

A. *Informant Interviews*

Structured face-to-face interviews of family members were conducted in 89% of the cases. The interview form contained 234 items concerning the victim's everyday life and behavior, family related factors, use of alcohol and drugs, previous suicidality, help seeking, and life events. The interviews took place about four months after the suicide and the mean duration of each was 2 hours 52 minutes (SD 65 minutes). Professionals were interviewed in 51% (n=59) of the cases. Structured face-to-face interviews of mostly psychiatric professionals who had attended the deceased during the preceding year were done in 22%, and semistructured interviews either face-to-face or by telephone with the health or social care professional the victim had last consulted before death in 49% of the cases. Supplementary unstructured interviews, most often by telephone, were made in 53% (n=61) of the victims. When all the data were collected, a multidisciplinary team discussed the cases, and case reports were written on the basis of all information available.

B. *Subjects*

There were 116 suicides, 19 females and 97 males, in the ages 13 to 22 years during the research period. Only six victims were under age 16 years. The mean age of the female victims was 19.1 years, SD=2.2 years, of the males 19.3 years, SD=2.1, and the male to female ratio was 5.1 to 1.

C. *Psychiatric Diagnoses and Level of Functioning*

Psychiatric diagnoses according to DSM-III-R (American Psychiatric Association, 1987) criteria were made in two phases. First, two psychiatrists independently made provisional best-estimate diagnoses weighing and integrating all available information from different sources. After this all cases with some kind of diagnostic disagreement were reanalyzed with a third psychiatrist to achieve general consensus for the final diagnoses. Multiple diagnoses on Axes I–III were made, when applicable (Henriksson et al. 1993; Marttunen et al. 1991).

The victims' psychological, social, and occupational functioning was assessed using the "Global Assessment of Functioning Scale" (GAF-Scale) of DSM-III-R (Axis V). During the final year the assessment was based on the highest level of functioning for at least a few months during the past year. For victims in school, this period included at least one month during the school year. The judgement of the current level of functioning was based on the period of one week prior to suicide. These assessments were made in consensus between two researchers.

D. *Statistical Methods*

Female and male suicides were compared using the Chi square test with Yates' correction, Fisher's exact probability test, Student's t-test, two-tailed, when appropriate.

TABLE 1

Suicide methods and use of alcohol in adolescent female and male suicides

Suicide methods and use of alcohol at time of suicide	Female suicides (N=19)		Male suicides (N=97)	
	N	%	N	%
Hanging	6	32	27	28
Overdose	5	26[1)]	8	8[1)]
Jumping from height	4	21[1)]	2	2[2)]
Shooting	1	5[3)]	42	43[3)]
Carbon monoxide	0	0	9	9
Drowning	1	5	1	1
Traffic	2	11	8	8
Suicide under influence of alcohol	9	47	58	60

1) $\chi^2 = 3.555$, df = 1, P = 0.059; 2) Fisher exact, P = 0.013; 3) Fisher exact, P = 0.002

II. Results

A. *Suicide Methods*

Shooting was less common but overdose and jumping from height more common suicide methods in female than in male suicides (Table 1). Suicide by violent methods (including hanging, shooting, cutting, jumping, traffic death) were common among both genders (13 females [68%], 79 males [81%]).

B. *Family Related Factors*

Irrespective of family constellation family problems – either parental psychopathology, violence, substance abuse or arguments between family members – were reported in 17 (89%) female victims. There were no statistically significant differences between females and males in terms of family background related factors (Table 2). Female victims were less often unemployed but more often incapable of working.

C. *Antisocial Behavior*

Antisocial behavior (including truancy, running away, recurrent stealing, use of illicit drugs, sexual promiscuity, violence against person, convictions and other difficulties with the law) (Marttunen et al. 1994) was twice as com-

TABLE 2

Family background and occupational status of adolescent female and male suicides

	Female suicides (N=19)		Male suicides (N=97)	
	N	%	N	%
Family background				
Parental divorce	7	37	29	30
Parental death	2	11	15	15
Parental alcohol abuse	8	42	24	25
Parental psychiatric disorder	3	15	20	21
Parental violence	5	26	15	15
Parental somatic illness	5	26	37	38
Parental suicide or suicide attempt	1	5	9	9
Low parental SES	12	63	55	57
Living with parent(s)	10	53	68	70
Institutional upbringing	1	5	9	9
Occupational status				
Student / in school	7	37	25	26
Working	6	32	44	45
Unemployed	0	0	15	16
Incapable of working	6	32[1)]	5	5[1)]

[1)] χ^2)=10.029, df=1, P = 0.002

TABLE 3

Antisocial behavior in adolescent female and male suicides

	Female suicides (N=19)		Male suicides (N=97)	
	N	%	N	%
Antisocial behavior lifetime	5	26[1)]	50	52[1)]
Antisocial behavior previous year	5	26	43	44
Violence against person lifetime	2	11	16	16
Violence against person previous year	2	11	16	16
Problems with the law	4	21	39	40
Convicted	1	5	17	18

[1)] $\chi^2 = 3.108$, df = 1, P = 0.078

TABLE 4

Previous suicidal behavior and suicide communication in adolescent female and male suicides

Previous suicidal behavior	Female suicides (N=19)		Male suicides (N=97)	
	N	%	N	%
Previous suicide attempt(s)	12	63[1)]	29	30[1)]
Two or more previous suicide attempts	8	42[2)]	9	9[2)]
First suicide attempt >3 months before suicide	8	42	24	25
Suicide talk	12	63	56	58
First suicide talk >3 months before suicide	8	42	25	26
Previous suicide attempt(s) or suicide talk	14	74	61	63

[1)] $\chi^2 = 6.305$, df = 1, P = 0.012; [2)] $\chi^2 = 11.190$, df = 1, * < 0.001

mon among males as among females (Table 3). Also convictions were more common among the male than female victims.

D. Previous Suicidal Behavior

Compared with male suicides the female victims had more often attempted suicide before the final attempt. Verbal communication of suicidal intent was relatively common both in the female and the male cases (Table 4). Most victims among both genders had communicated their suicidal intent either by suicide attempts or by talking about suicide.

E. Psychiatric Diagnoses and Psychosocial Impairment

All the female victims with sufficient data for diagnostic assessment received at least one psychiatric diagnosis, whereas there were 8 young males with no psychiatric diagnosis (Table 5). Adjustment disorders were three times more common among male than among female victims. Abuse or dependence of alcohol or other substances were almost as common in female as in male cases. Major depression was more common in female suicides. Female victims also more often received a diagnosis of borderline personality disorder (BPD). All the female victims with BPD had a comorbid mood disorder or alcohol abuse. Psychiatric comorbidity was found equally often among both genders.

TABLE 5

Selected DSM-III-R Axis I and II diagnoses in adolescent female and male suicides

Psychiatric diagnosis	Female suicides (N=19)		Male suicides (N=97)	
	N	%	N	%
No psychiatric diagnosis	0	0	8	8
Depressive syndrome[a]	15	79	54	56
Major depression	7	37	14	14[1]
Adjustment disorder	1	5	16	17
Substance abuse[b]	4	21	25	26
Non-affective psychotic disorders	2	11	7	7
Personality disorders	6	32	33	34
BPD[c]	5	26	7	7[2]
Antisocial disorders[d]	1	5	11	11

[a] includes mood disorders and adjustment disorder with depressed mood
[b] includes also victims with a diagnosis of substance dependence
[c] BPD indicates borderline personality disorder
[d] includes antisocial personality disorder and conduct disorder
[1] $\chi^2 = 3.976$, df = 1, P = 0.046; [2] $\chi^2 = 4.359$, df = 1, P = 0.036

Psychosocial impairment during the week preceding suicide was more severe among the female than among the male victims (mean GAF score 37.9, SD=21.0 for females vs 50. 9, SD = 21.1 for males, t = 2. 45, df= I 14, $P < 0.02$). The difference between the genders in psychosocial impairment during the year before the suicide was not statistically significant (mean GAF score 59.7, SD=22.1 for females vs 65.1, SD=22.1 for males).

F. Previous Psychiatric Care

The female victims had more often been in psychiatric care during the year preceding suicide than the males (Table 6). History of inpatient psychiatric care was also more common among female victims.

III. Discussion

The high rate of family problems among both male and female adolescent suicides is consonant with the results by Brent et al. (1994a) reporting familial psychopathology and parent-child conflict to be related to suicide risk in adolescence.

TABLE 6

Contacts with psychiatric and other health care

Contacts with psychiatric and other health care	Female suicides (N=19)		Male suicides (N=97)	
	N	%	N	%
Lifetime				
Psychiatric care	9	47	31	32
Inpatient psychiatric care	8	42[1)]	12	12[1)]
Previous year				
Psychiatric care	9	47[2)]	20	21[2)]
Inpatient psychiatric care	8	42[3)]	8	8[3)]
Previous month				
Psychiatric care	9	47[4)]	7	7[4)]
Inpatient psychiatric care	8	42[5)]	3	3[5)]

1) $\chi^2 = 7.871$, df = 1, P = 0.005; 2) $\chi^2 = 4.721$, df = 1, P = 0.03;
3) $\chi^2 = 12.602$, df = 1, P < 0.001 4) $\chi^2 = 18.297$, df = 1, P < 0.001;
5) Fisher exact, P < 0.001

Consistently with previous studies using the psychological autopsy (Gould et al. 1992) and studies based on coroners' data (Cosand et al. 1982; Hoberman & Garfinkel, 1988), previous suicide attempts were more prevalent among female than male suicides. This gender difference may partly be a consequence of the higher rate of suicide attempts among adolescent females in the general population (Ostamo & Lönnqvist, 1994; Shaffer et al. 1988), and also reflect the choice of less lethal methods by females in their suicide attempts resulting to a higher frequency of nonfatal attempts.

As previously reported (Shaffer et al. 1994) in suicides under age 20, major depression was more common in young female than male suicides also in the present study population including victims aged 20 through 22 years. Adjustment disorders were three times more common among the male victims. No females but 8% of the males received no psychiatric diagnosis. However, probably due to the relatively small number of female cases, the differences did not reach statistical significance. The finding that antisocial behaviors and violence were more common among male than female suicides underscores the need to evaluate risk for suicide among (male) adolescents with antisocial tendencies.

Unexpectedly, the frequency of substance abuse or dependence among the female victims was almost as high as among the males. Another unexpected result was the high proportion of female victims completing suicide under the influence of alcohol. These findings indicate the importance of alcohol related problems in youth suicides among both genders. Brent et al. (1994b) reported that personality disorders are critical risk factors for adolescent suicide but they did not report gender differences. Different from the results of Runeson and Beskow (1991) who found no gender difference in the prevalence of BPD in adolescent and young adult suicides, BPD and the combined group of BPD and identity disorder were more common among the female than male suicides in the present study. Previous studies specifically on adolescent suicide have not reported the frequency of BPD among female victims or compared the genders.

The frequency of female victims with psychiatric care during the year preceding suicide was comparable to previous research (Thompson, 1987; Asgard, 1990). The high frequency of inpatient care among the females is likely to indicate that the young female victims had suffered from more severe psychiatric disorders than the young males. The more severe psychosocial impairment among the females further supports this possibility.

The results suggest that young females committing suicide may have suffered from more severe psychopathology than young male suicides. However, due to the small number of female suicides in this study, one must draw conclusions with caution. Further, the risk ratios associated with specific mental disorders cannot be assessed on the basis of the study design. The need for further research on the specific characteristics of suicides among female youth and gender differences in youth suicide with large enough representative suicide populations is obvious.

Early detection and effective treatment of adolescents suffering from major mood disorders probably is an effective prophylactic intervention against suicide (Ryan et al. 1987). From the viewpoint of suicide prevention, there clearly is a need to develop effective treatments for adolescent depressive disorders, particularly for major depression. Given the higher prevalence of major depression and psychiatric treatment contacts prior to suicide among adolescent females than males this prevention strategy is likely to be relatively more beneficial for females than males. In general, preventive efforts within psychiatric care are likely to reach a higher proportion of young females than males at high risk for suicide suggesting that different strategies may be needed in the prevention of female and male adolescent suicides.

Note

*Partly based on article: Marttunen, M., Henriksson, M., Aro, H., Heikkinen, M., Isometsä, E., Lönnqvist, J.: "Suicide among female adolescents. Characteristics and comparison with males in the age group 13 to 22 years". J. Am. Acad. Child Adolesc. Psychiatry 1995 34 (in press).

References

American Psychiatric Association. Diagnostic and Statistical Manual of Mental Disorders. 3rd edition revised (DSM-III-R). Washington, DC: American Psychiatric Association, 1987.

Apter A, Bleich A, King, RA, Kron S, Fluch A, Kotler M, Cohen DJ. Death without warning? A clinical postmortem study of suicide in 43 Israeli adolescent males. Arch Gen Psychiatry, 1993; 50: 138–142.

Åsgård U. A psychiatric study of suicide among urban Swedish women. Acta Psychiatr Scand, 1990; 82: 115–124.

Barraclough B. International variation in suicide rate of 15–24 year olds. Soc Psychiatry Psychiatr Epidemiol, 1988; 23: 75–84.

Brent DA, Perper JA, Goldstein CE, Kolko DJ, Allan MJ, Allman CJ, Zelenak JP. Risk factors for adolescent suicide. A comparison of adolescent suicide victims with suicidal inpatients. Arch Gen Psychiatry, 1988; 45: 581–588.

Brent DA, Perper JA, Moritz G, Liotus L, Schweers J, Balach L, Roth C. Familial risk factors for adolescent suicide: a case-control study. Acta Psychiatr Scand, 1994a; 89: 52–58.

Brent DA, Johnsos BA, Perper J, Connolly J, Bridge J, Bartle S, Rather C. Personality disorder, personality traits, impulsive violence, and completed suicide in adolescents. J Am Acad Child Adolesc Psychiatry, 1994b; 33: 1080–1086.

Cosand BJ, Bourque LB, Kraus JF. Suicide among adolescents in Sacramento county, California 1950–1979. Adolescence, 1982; 17: 917–930.

Gould M, Shaffer D, Fisher P, Kleinman M, Morishima A. The clinical prediction of adolescent suicide. In: Assessment and prediction of suicide, Maris R, Berman A, Maltsberger J, Yufit R (editors). New York: Guilford Press, 1992: 130–143.

Henriksson M, Aro H, Marttunen M, Heikkinen M, Isometsä E, Kuoppasalmi K, Lönnqvist J. Mental disorders and comorbidity in suicide. Am J Psychiatry, 1993; 150: 935–940.

Hoberman HM, Garfinkel BD. Completed suicide in children and adolescents. J Am Acad Child Adolesc Psychiatry, 1988; 27: 689–695.

Lesage A, Boyer R, Grunberg F, Vanier C, Morissette R, Menard-Buteau C, Loyer M. Suicide and mental disorders: A case-control study of young men. Am J Psychiatry, 1994; 151: 1063–1068.

Lönnqvist J. National suicide prevention project in Finland: a research phase of the project. Psychiatria Fennica, 1988; 19: 125–132.

Marttunen, MJ, Aro HM, Henriksson MM, Lönnqvist JK. Mental disorder in adolescent suicide. DSM-IIIR axes I and II among 13 to 19 year olds. Arch. Gen. Psychiatry, 1991; 48: 834–839.

Marttunen MM, Aro HM, Lönnqvist JK. Adolescent suicide: endpoint of long-term difficulties. J Am Acad Child Adolesc Psychiatry, 1992; 31: 649–654.

Marttunen M, Aro H, Henriksson M, Lönnqvist J. Antisocial behaviour in adolescent suicide. Acta Psychiatr Scand, 1994; 89: 167–173.

Ostamo A, Lönnqvist J. The epidemiology of attempted suicide in Helsinki, 1989–1993. In: Attempted suicide in Europe. Findings from the multicentre study on parasuicide by the WHO regional office for Europe, Kerkhof A, Schmidtke A, Bille-Brahe U, de Leo D, Lönnqvist J (editors). Leiden, the Netherlands: DSWO Press, 1994: 137–158.

Rich CL, Young D, Fowler RC. San Diego suicide study I: young vs old subjects. Arch Gen Psychiatry, 1986; 43: 577–582.

Runeson B. Mental disorders in youth suicide. DSM-III-R axes I and II. Acta Psychiatr Scand, 1989; 79: 490–497.

Runeson B, Beskow J. Borderline personality disorder in young Swedish suicides. J Nerv Ment Dis, 1991; 179: 153–156.

Ryan ND, Puig-Antich J, Ambrosini P, Rabinovich H, Robinson D, Nelson B, Lyengar S, Twomey J. The clinical picture of major depression in children and adolescents. Arch Gen Psychiatry, 1987; 44: 854–861.

Shaffer D. The epidemiology of teen suicide: An examination of risk factors. J Clin Psychiatry, 1988; 49, 9(suppl): 36–41.
Shaffer D, Garland A, Gould M, Fisher P, Trautman P. Preventing teenage suicide: a critical review. J Am Acad Child Adolesc Psychiatry, 1988; 27: 675–687.
Shaffer D, Gould M, Hicks R. Epidemiology, mechanisms and clinical features of youth suicide. In: Divergent perspectives on suicidal behaviour, Kelleher M (editor). Fifth European Symposium on Suicide, Cork, 1994: 79–95.
Shafii M, Carrigan S, Whittinghill JR, Derrick A. Psychological autopsy of completed suicide in children and adolescents. Am J Psychiatry, 1985; 142: 1061–1064.
Thompson TR. Childhood and adolescent suicide in Manitoba: A demographic study. Can J Psychiatry, 1987; 32: 264–269.
World Health Organization. World Health Statistics Annual 1991. Geneva, Switzerland: World Health Organization, 1992a.
World Health Organization. World Health Statistics Annual 1992. Geneva, Switzerland: World Health Organization, 1992b.
World Health Organization. World Health Statistics Annual 1993. Geneva, Switzerland: World Health Organization, 1993.

11. The Suicidal Process in Young Suicides

BO S. RUNESON

I. Introduction

How long does it take from the first thought on suicide until a young person finally commits suicide? What factors influence the suicidal process?

We have previously reported on the prevalence of mental disorder (DSM-III-R) in a series of consecutive suicides in the ages 15- to 29-year-olds from Göteborg, Sweden. This presentation applies another perspective, focusing on the suicidal process, i.e. the progression of suicidality within the individual in interaction with surrounding people [1].

Did they have models for their behaviour? Did they communicate their wishes? How did they choose a certain method for suicide? How was the suicide performed?

II. Material and Method

The data presented are based on modified psychological autopsies, where we need informants with an intimate relationship to the suicide victim. In these 58 suicides (of 42 men and 16 women), I was allowed to interview one or more members of the family of origin in 86%. In the other cases I talked with a foster mother, girlfriends, close mates etc. I made supplementary interviews with personnel from school or work, with staff from health care or social welfare. The data were of a current character, in 93% the last contact of the main informant was within the last two weeks before death.

Still, we are aware of the problems in the retrospective study design. One has to manage the biases of both the bereaved informants and the investigators [2].

The main diagnostic groups were major depression (n=13), schizophrenia (n=8), adjustment disorder (n=8) and borderline personality disorder, BPD, (n=16). The remaining 13 subjects had miscellaneous diagnoses: anorexia nervosa, alcohol dependence etc. (3–4).

D. De Leo, A. Schmidtke and R.F.W. Diekstra (eds.), Suicide Prevention, 105–110.

TABLE 1

Exposure to suicidal model in parents/siblings in relationship to sex and principal diagnosis of victims

	Suicidal behaviour (%)	Death (%)
Men	40	19
Women	31	–
Major depression	31	15
Schizophrenia	63	13
Adjustment disorder	13	–
Borderline personality disorder	56	25

III. Results

A. Exposure to Suicidal Models

40% of the young men in this series had been exposed to a parent or sibling who had tried to commit suicide (Table 1). Few had actually committed suicide, only one mother and two older brothers of three male suicide victims died by suicide. Other deaths were more common, several of which were alcohol-related and occurred in fathers of suicide victims.

B. Current Warnings

A verbal message or sometimes a suicidal gesture or attempt was evident in 78%, sometimes during the weeks or years preceding the suicide, i.e. ever. An open clearly expressed suicidal communication *in connection with the current episode* occurred in only 41% of the cases. A lack of any communication was slightly more common in males ($P = 0.06$), and in subjects with an adjustment disorder or major depression ($P < 0.001$).

C. Previous Suicide Attempts

Previous attempts were common in both sexes, 60% of the men and 81% of the women had made at least one attempt. Repeated attempts were more common in women. More than 2 attempts were found in 56% of the women and in 17% of the men ($P < 0.01$). These figures are higher than in other studies of youth suicide.

One mother said that she "had stopped counting at ten (suicide attempts)". Another mother said that "from the second suicide attempt, I never dared to talk about it with my daughter again". Repeated suicide attempts were more

TABLE 2

Method of suicide attempts and suicide

18/38 (47%) escalated from poisoning/cutting to active methods
13/38 (34%) used methods of varying severity
6/38 (16%) used the same method in all attempts and suicide

common among young women, not unexpectedly. Suicide attempts existed in all diagnostic groups, however, more seldom in adjustment disorder – and had in these cases often taken place only a day or two before the suicide. Repeated attempts were significantly less common in this diagnostic group as well.

D. Method of Suicide Attempts and Suicide

The methods of previous attempts were in 63% poisoning and secondly cuttings. The methods for the suicide were among men hanging followed by poisoning and in women the same methods in the opposite order.

The choice of method for attempt and suicide was very rarely influenced by the parent's method. Nor did they continue to use one certain method consistently, that was evident in only 16% (Table 2). They seemed to experiment with different methods, which also reflects the stronger intention to die later in the process. An escalation was found in a large share. They seemed to learn by their own practicing [5].

Many of the suicide attempts were of seemingly low intention to die. Only 10 of 35 had >8/16 points on Beck's Suicidal Intention Scale, objective circumstances, when evaluated for the cases where data were sufficient concerning the previous suicidal attempt. We conclude that one has to evaluate even communicative suicidal acts as a possible predictor for suicide.

E. Suicide Notes

A written message related to the suicide was found postmortem in 38%. The messages were typically short and contained a plea of forgiveness. In a minority the content was negative against the persons addressed.

F. Toxicologic Screening

Postmortem toxicology was positive for alcohol in 33% and benzodiazepines in 24%. Benzodiazepines were less common in suicide victims with adjustment disorder or major depression (P = 0.05). The occurrence was similar concerning alcohol, but not statistically significant.

TABLE 3

Duration of suicidal process

	Median time (months)	<2 months (%)	>2 years (%)
Major depression	3	54	31
Schizophrenia	47	0	100
Adjustment disorder	0	75	0
Borderline personality disorder	30	0	63

G. Duration of Suicidal Process

The time that elapsed from first observed indication of suicidal ideation to the completed suicide, i.e., the duration of the process from the first communication or sometimes attempt, is shown in Table 3. The median time is given in months, frequencies of short (<2 months) or long (>2 years) duration are given in per cent.

The median time for men was 12 months and for women 42 months (P = 0.02). A short process was most common in suicide victims with a principal diagnosis of adjustment disorder and major depression. A longer process was more common in BPD and especially schizophrenia. When a survival analysis according to Breslow (Generalized Wilcoxon) was applied, the differences between diagnostic groups were found statistically significant.

It was more difficult to define the starting point of the mental disorder. Probably because this was less dramatic than the start of the suicidal behaviour. However, among the 13 subjects with primary major depression, 7 had a recurrent episode and 6 had a first episode of depression. When assessing the severity of the prevalent depression, 5/13 had severe symptoms, 5/13 had moderate symptoms and 3/13 had mild symptoms of were in partial remission.

H. Psychosocial Stressors

Severity of the stressors precipitating the suicide was assessed in accordance with DSM-III-R Axis IV (Table 4). Levels 2–3 means mild or moderate severity, 4–5 indicates severe to extreme stressors. We found that among schizophrenics the stressors were extreme and enduring, while e.g. in adjustment disorder and also often in BPD, the stressors showed a tendency of being milder.

TABLE 4

Level of severity of stressors, DSM-III-R Axis IV

	2–3	4–5	P
Major depression	5/13	8/13	n.s.
Schizophrenia	0/8	8/8	0.0001
Adjustment disorder	6/8	2/8	0.07
Borderline personality disorder	10/16	6/16	n.s.

I. Witnessed Suicides

The suicide was performed in front of someone in 14/58 subjects. In 5 cases the observer was a significant person, sometimes there was an obvious intention to harm the one observing the death. This aggressive nature was more common among subjects with BPD than among victims with other diagnoses 7/16 (44%) vs 7/42 (17%) ($P < 0.05$).

IV. Conclusions

The processes were even in young adults often several years long. Previous suicide attempts were found in 2/3, often of seemingly low intention. Suicidal communication close to the suicide was fairly uncommon.

- The suicidal process in major depression was fairly short, communication and drug poisoning at the final act uncommon. First-time/recurrent depressions were equally common, the severity of the symptoms was not necessarily high.
- The previous suicidal behaviour in adjustment disorder had the shortest duration, previous attempts were uncommon. Psychosocial stressors were mild. These deaths were totally unexpected and took place among previously healthy subjects.
- The suicidal processes in BPD were fairly long, models for suicidal behaviour within the family were frequent. Witnessed suicides were typical, as was drug poisoning at the final act.
- The schizophrenic suicidal processes were long, suicidal models were common. Psychosocial stressors were severe and enduring.

References

1. Beskow J. Longitudinal and transectional perspectives of suicidal behaviour. Experiences of suicide prevention in Sweden. Suicide Research II. Psychiatr Fennica suppl, 1983: 55–64.
2. Beskow J, Runeson B, Åsgård U. Psychological autopsies: Methods and ethics. Suicide Life Threat Behav, 1990; 20: 307–323.
3. Runeson B. Mental disorder in youth suicide. DSM-III-R Axes I and II. Acta Psychiatr Scand, 1989; 79: 490–497.
4. Runeson B, Beskow J. Borderline personality disorder in young Swedish suicides. J Nerv Ment Dis, 1991; 179: 153–156.
5. Diekstra RFW. A social learning theory approach to the prediction of suicidal behaviour. Proceedings 7th International congress on suicide prevention. Swets & Zeitlinger, Amsterdam, 1974: 55–66.

12. Childhood Conceptions of Death and Suicide

Empirical Investigations and Implications for Suicide Prevention

BRIAN L. MISHARA

When do children begin to form an understanding about suicide? What do they know and think about suicide at different stages of development? How do their conceptions of suicide develop and change? How do children learn about suicide? What are the implications of children's understanding of suicide for suicide prevention? The above questions are the major focus of this paper.

Several years ago, a grant proposal which the author of this paper submitted for funding was rejected with the following comment: "It is a waste of valuable resources to support the study of suicide in children since children almost never kill themselves, and those rare cases when children are thought to commit suicide may be more accurately be described as accidental deaths because children cannot understand the finality and seriousness of their acts." These comments are particularly interesting because they raise questions which may be verified by empirical research: what do children actually understand and know about death and suicide? But it is even more interesting that the author of these comments ignored the possibility that knowledge about how conceptions of suicide develop in childhood may help in the prevention of suicidal behaviour in adolescents.

Despite the lack of funding for this research over the past years, I have been involved in asking children aged 6 to 12, from the first to the sixth grades, in Canada, about suicide, life and death. We asked questions about the concept of life, based upon Piaget (1937). "What does it mean to be alive? Is the sun alive? Why? Or why not? Is fire alive? What about a mountain etc?" We asked questions about children's understanding of death based upon previous studies (e.g. Grenier, 1986; Normand & Mishara, 1992; Lonetto, 1980; Carlson, Asarnow & Orbach, 1987; Koocher, 1973). For example, we asked: "What does it mean to be dead? What happens when we die? What happens after we die? Can we see? Can we hear? How does one feel? Why do people die?"

We also asked many questions about suicide. We asked: "What does suicide mean? What is the difference between dying and "suiciding"? (We can ask

D. De Leo, A. Schmidtke and R.F.W. Diekstra (eds.), Suicide Prevention, 111–119.

this question because these were French-Speaking children and "suiciding" (se suicider) exists as a verb in French). When someone is very sick and dies, is this a suicide? Why do you say that? If someone is hit by a car while crossing the street, is this a suicide? If someone drinks poison? How can people suicide? Why do some people want to suicide? What happens afterwards? Do you think an animal can suicide? Why? What about someone your age? Have you ever thought about suicide?"

We asked about children's experiences with death and suicide, as well as questions from tests on cognitive development in order to determine the relationship between the level of cognitive development and their level of sophistication in their understanding of death and suicide. We asked if anyone ever talked to them about suicide, and if they saw or heard about suicide from films or television. Did they know anyone who committed suicide? We also asked: "What should you say to someone who says to you he or she is thinking of suiciding?"

The interviews were quite lengthy and most children enjoyed talking with us. Only one child out of 135 children in two separate studies seemed upset about discussing suicide during the interview, and this child had recently lost a relative by suicide (I wonder if the mother who approved of our interview on the topic expected us to explain something to the child which she had not yet discussed, but that the eight-year old told us he knew "all about.")

Our studies were not the first to investigate the topic of children and suicide. Official statistics in Canada (Suicide in Canada, 1994) and the United-States (National Center for Health Statistics, 1988) indicate that, according to official statistics children rarely commit suicide. Between 1950 and 1992 in Canada, not a single child under age 5 was recorded as having committed suicide. During the same period, there were only 18 reported suicides between 1950 and 1992 in children aged 5 to 9, which results in a suicide rate of .2 deaths per 100,000 population per year. In Canada, between 1951 and 1992, suicide rates in 10 to 14 year olds increased from 0.1 per 100,000 population per year (only 1 death) to 1.8 per 100,000 in 1992 (34 deaths). This rate of 1.8 compares with the rate of 12.9 per 100,000 per annum for the 15 to 19 year old age group.

Several researchers have suggested that the official statistics on children's suicides significantly underestimate the actual frequency of the phenomenon for younger ages (Cohen, Sandler, Berman & King, 1982; Hoberman & Garfinkel, 1988; Matter & Matter, 1984; Pfeffer, Lipkins, Plutchik & Mizruchi, 1988). For example, Pfeffer, Lipkins, Plutchik and Mizruchi (1988) found that 2% of a sample of pre-adolescents with no psychiatric history had made suicide threats, 1% had made a "mild" suicide attempt and 8.9% had thought of suicide. Other authors point to the fact that accidents are the number one cause of children's death in Canada and the United-States, with the most common type of accidental death being children hit by cars. Winn and Heller (cited in Shaffer & Fisher, 1981) observed that the majority of children who threatened to commit suicide were planning to die by throwing themselves in front of a moving car. Normand and Mishara (1992) found that running in front of a car was often reported by children as a possible means for people to commit suicide. Shaffer and Fisher (1981) observed that there were no reports of any child who committed suicide by being hit by a car in the United Kingdom. Hoberman and Garfinkel (1988) estimated that 15% of deaths which were actually by suicide in people under age 20 in the United-States had been classified as deaths by "accident", "homicide" or "undetermined", and that this misclassification occurred more often in those below age 15.

A few studies have asked children if they had thought of suicide. Pfeffer asked 101 normal school age children between the age of 6 and 12 if they have ever thought of killing themselves. She found that 8.9% said they had thought of killing themselves and another 2% said they threatened to commit suicide or that they had attempted suicide. Carlson, Asarnow and Orbach (1987) reported that 15.4% of normal children between age thirteen and eighteen had suicidal fantasies.

Although there have been relatively few studies on children's conceptions of suicide there has been greater interest in the development of children's understanding of death. In the 1940s, probably as a result of the War, Sylvia Anthony (1940) and Maria Nagy (1948) interviewed children concerning their concept of death and reported a series of stages in the acquisition of a "mature" understanding of death. For the youngest children, death was akin to a sleep-like state, from which one can be awakened, as in the fairy tale Sleeping Beauty. For the youngest children it was possible to avoid death by being adept or careful and dead people were sometimes considered to have thoughts, feelings and behaviours which more mature children ascribed only to the living.

Since this early research, there has been much debate over the exact nature of the stages in the acquisition of a mature concept of death. For example, Melear (1973) found that some younger children had a more "mature" understanding than several of the older children. Raimbault (1975) reported that children who suffer from a terminal illness develop a sophisticated understanding of death, even at a young age. Koocher (1973) found that the concept of death was related to levels of cognitive development, classified according to a Piagetian schema, rather than chronological age. Regardless of the debates concerning the possibility of defining clear cut stages, researchers generally agree that an understanding of death develops gradually. More recent research has evaluated the concept of death according to five sub-concepts: finality, irreversibility, causality, universality and aging (e.g. Orbach, Gross, Glaubmann & Berman, 1985). Aging, the idea that people die eventually when they grow old, tends to be understood before any other concept. Irreversibility, the idea that once a person is dead they cannot come back to life, is usually understood next, followed by universality (everyone dies). Finality (the idea that death is an end state) and causality (the concept that death is an inevitable part of living and is not just due to some external accident) develop later. Grenier (1986) refined these sub-concepts and developed criteria for a "mature" concept of death which were: finality (which includes irreversibility), "state of death" (which concerns how complete is the cessation of biological functions), universality, unpredictability and causality.

Grenier (1986) found that the more experiences children had related to death (e.g. prolonged separations, death of an animal or pet, death education at school, etc.) the more mature their concept of death in general. In her study of 52 six to nine year olds, Kane (1975) found that death concepts tended to be more mature when children experienced a death in the immediate family and younger children (age three to six) tended to change their concept of death the most as a result of the experience of a loss of a family member by death. Overall, our review of the literature suggests that the concept of death develops gradually and is related to experiences with death. One can ask if the concept of suicide develops in a similar manner and more particularly when do children begin to understand about suicide and how does their understanding relate to their comprehension of death as well as their experiences with death and suicide.

Normand and Mishara (1992) conducted an interview study of 60 French-speaking public elementary school children in Montreal, Canada, in grades 1, 3 and 5. They evaluated the children's understanding of death, experiences related to death, as well as their understanding of the word "suicide", experiences related to suicide and attitudes towards suicide. In this study school personnel felt that if a child did not know the meaning of the word, "suicide", that the interview with these children should be stopped immediately rather than pursuing the investigation of the child's understanding of words such as "ending one's life" or "killing oneself." In that study the researchers ended the interview without further investigation if the child did not know the meaning of the word, "suicide." Only one of the 60 children chose to terminate the interview before its completion because he felt "too sad to talk about death" (his mother had died a few months earlier). Otherwise, children generally discussed death and suicide easily with no negative emotional reactions.

The study by Normand and Mishara (1992) found that 87% of the children understood the concept of the universality of death, that is, that everyone will die some day, even without having an accident or contracting an illness. Almost 90% of the children understood that death is final, that it is impossible to come back to life. Slightly more than half of the grade school children (53%) recognized that death is unpredictable. Ninety percent of the children reported causes of death which were external, such as accidents or smoking cigarettes. Only 10% cited internal processes as being the cause of death. Forty percent of the children felt that people who were dead could feel emotions such as "good" or "sad" even though they could not see or move and one third felt that people who died continued to live in some manner in another setting,

Heaven. In this study children had a mature concept of death by the age of 9 or 10.

In the study by Normand and Mishara, only two of the twenty first graders knew the meaning of the word "suicide" and only fifty percent of the third graders knew what "suicide" meant. However, nineteen of the twenty fifth graders knew the meaning of the word. Because of the possibility that children may understand about killing oneself but not know the word, "suicide", Mishara (1995) conducted a follow-up study which differed from the study by Normand and Mishara (1992): children who did not understand the meaning of "suicide" or "to commit suicide" (in French: "se suicider") were then asked if they knew about "killing oneself." Children who knew about "killing oneself" but did not recognize the word "suicide" were then interviewed in detail about suicide, replacing "killing oneself" for "suicide" throughout the interview. This study involved 65 children in each of the grades from one through five. Although none of the six year olds knew about suicide in this sample, one third of seven year olds knew about suicide or killing oneself, 87% of eight year olds and 81% of nine year olds as well as all (100%) of children age ten and older. Five of the sixty-five children (7.7%) reported that they knew someone who committed suicide and nine of the children (13.8%) had considered killing themselves.

On the basis of the research by Mishara (1995), Normand and Mishara (1995) and others we can conclude that: children, particularly by about age eight, understand what suicide is and are able to talk about it in an interview, including presenting their explanations of why people kill themselves. Children may also be aware when someone they know personally commits suicide (although in no cases did parents tell any of the children about the suicide) and some children have thought about killing themselves.

Although virtually all children know about death, their understanding of death differs from what adults generally think. For the youngest children, death is not necessarily final. Death is seen as similar to what children see in television cartoons: the characters may be crushed to the thickness of a piece of paper but then they can jump up and come back to life. Death is not inevitable: If we have a good doctor or are careful (e.g. crossing the street carefully or avoiding bad guys), we can avoid dying. Everyone does not necessarily die – death is not universal; and death is seen as being caused only by exterior events. Also, death is not exactly like the death adults envision. Consider this interview conversation:

> "Can dead people see?" – "No."
> "Why not?" – "Their eyes are closed."
> "If they opened their eyes, could they see?" – "No, because it is dark in the coffin. But if they had a flashlight with them, they could see perfectly."

Many of the attributes which we reserve for the living, such as feeling sad, are ascribed to the dead by younger children. But, more important is the fact that children can often accept the co-existence of different, seemingly contradictory, concepts of death without any trouble. A child can understand that everyone dies some day, and at the same time believe that, if you are careful, you may live forever. A young child may believe that people who die can never come back to life, that death is final, – but if you want to come back *very much* and your friends shake you *very hard,* you can wake up from death. All this suggests that suicide may have a very different meaning for a child for whom death is not a complete cessation of living from which one can never return.

A child's understanding of suicide may also have a significant influence, not only upon the child's own potential for suicidal behavior, but also on the child's overall adjustment and development throughout childhood and adolescence. To illustrate this, consider the case of an eight-year-old boy whose presenting problem was that of having an eating disorder. The child was much below the average weight for his height and age. He was reported as giving away the contents of his lunch box every day and his parents said he hid food from meals rather than eating, and gave this food to other children in the neighbourhood or at school. Although the child lived in a poor neighbourhood where nutrition was not generally good, he came from a "better off" family who were proud that they were able to provide quality food for their child. In discussions with the child, he said that he gave food to the other children because they *wanted* his food. Further investigation revealed that when he was four years old his older half-brother (then 16 years old) committed suicide by hanging himself. It was this young boy who had discovered the body. The parents had told the child that this was an accident. However, the child remembered clearly that his older brother had asked his father to borrow the car that day and was refused shortly before killing himself. The boy, four years later, was afraid that if he did not give his friends his food, which they had asked for, something as disastrous might happen to them, in the same way that his half-brother had killed himself because his father did not give him something he had wanted. The child even went so far as to generalize this behavior and not wait for the others to ask – he insisted upon giving them his food as often as possible, whether they asked or not.

In this instance, a young child reached his own understanding of why a tragic loss had occurred in his family in a situation where no one explained what had really happened. Four years later this understanding was expressed by what was erroneously diagnosed as an eating disorder. The child responded well to therapy, whose goal was to help the child develop a more mature (and perhaps more sophisticated) understanding of his brother's death and why people may kill themselves.

In the study by Mishara (1995) children learned about suicide, primarily from discussion with other children and through experiences with suicide depicted on television. Three-quarters of the seven and eight year olds had seen or learned about suicide on television. All (100%) of the nine, ten, eleven and twelve year olds said that they had seen or learned about suicide from television. In this study children often reported talking about suicide – over 80% of children had discussed suicide. However, their discussions were usually with other children. It was rare that a parent or teacher would explain suicide to them.

It is frightfully naive to think that children do not know about suicide. Not only do they know about suicide, they also develop, moreover, an understanding of why and how people kill themselves, and this understanding is a function of their level of development and their experiences with death and suicide. In our society, and in all societies that the author is aware of, we rarely teach children about suicide and tend not to discuss suicide openly. Even when there is a death by suicide in the family, the child is usually told nothing or given an inaccurate explanation, such as, that an accident has occurred. However, even though explanations of suicidal death are not given, children often know what happened and find their own ways of understanding why the person committed suicide. Their explanations may be based upon personal hypotheses, discussions with others (generally other children) or from experiences with suicide in television or movies.

In television programs which children watch, suicides are often depicted as an heroic act which must be carried out when a fictional character is cornered or without a means of escape. There are also cartoon examples of mock suicides, where a character uses a suicide threat to obtain an unreasonable request. For example, a Bugs Bunny cartoon showed the rabbit threatening to shoot himself if another cartoon character did not hand him a carrot he desired. This suicide threat was successful in obtaining the carrot, although the rabbit showed afterwards that it was only a toy gun with which he could not have killed himself.

All of the children who understood about suicide were able to name at least one means of killing oneself; most mentioned several methods. The most frequently mentioned responses to the question "How can someone commit suicide?" involved using a knife. Thirty-eight of the 65 children (58%) mentioned using a knife as a possible method. Jumping was mentioned by 22 of the children (34%), using a firearm by 20 children (31%), poison by 16 (25%), intentionally being hit by a car by 8 children (12%), drowning by 6, hanging by 4, setting oneself on fire in 3 instances and one child mentioned banging one's head against a wall. These methods are for the most part realistic means of killing oneself and many of the methods are readily accessible to children. Accidents are the number one cause of death among children in Canada, although it is rare that accidental deaths by methods mentioned, such as falling from a building or being hit by a car, accidental burning or drowning, are considered suicides.

In the light of these research findings on children's understanding of suicide and death, it is evident that children acquire an understanding of suicide at an early age despite the fact that the topic is rarely discussed. Rather than allow children to develop their knowledge based upon hearsay or media depictions of suicide as heroic, it is suggested that it may be appropriate to develop ways of teaching children about suicide at an early age. Based upon the research findings, it is unlikely that children by age seven or eight have not already developed elaborate conceptions of suicide and death. Perhaps teaching efforts could begin with a discussion of children's own conceptions of suicide and death, and if appropriate, alternative interpretations of suicide and death may be presented. This may prove to be a promising method of primary prevention of suicidal behaviour later in life. What children learn about suicide at a young age develops and becomes transformed into their teenage and adult conceptions of suicide. If we do not teach children about suicide, they will learn themselves early in life, most probably from violent depictions of suicide in the media or incomplete and possibly erroneous explanations given by other children. It would also be warranted to take a close look at how suicides are depicted in television, particularly glorification of some suicides as the "only alternative," when cornered. These implausible depictions may present undesirable models for children who may be susceptible to identify with the characters.

Note

This paper is a shorter version of an original paper in: "Acta Psychatrica Scandinavica", 1995 (in press). It is printed with the permission of Munksgaard International Publishers Ltd, Copenhagen.

References

Anthony S. The Child's Discovery of Death: A Study in Child Psychology. London: Kegan, Paul, Trench, Trubner & Co., 1940.

Carlson GA, Asarnow JR, Orbach, I. Developmental aspects of suicidal behavior in children, Journal of the Psychiatry, 1987; 26: 186–192.

Cohen-Sandler R, Berman AL, King R. Life Stress and Symptomatology: Determinants of Suicidal Behavior in Children, Journal of the American Academy of Child Psychiatry, 1982; 21: 178–186.

Grenier G. L'acquisition d'un concept de mort évolué chez l'enfant en fonction des expériences vécues d'une part et du développement des concepts de vie, d'inclusion et d'age d'autre part. Unpublished Masters thesis, Université du Québec a Montréal, Montréal, Canada, 1986.

Health Canada. Suicide In Canada: Update of the Report of the Task Force on Suicide In Canada. Ottawa: Minister of Health, 1994.

Hoberman HM, Garfinkel BD. Completed suicide in children and adolescents, Journal of the American Academy of Child and Adolescent Psychiatry, 1988; 27: 689–695.

Kane R. Children's concept of death. Unpublished Doctoral Thesis, Department of Psychology, University of Cincinnati, 1975.

Koocher GP. Childhood, death and cognitive development, Developmental Psychology, 1973; 9: 369–375.

Lonetto R. Children's Conceptions of Death. New York: Springer Publishing Co., 1980.
Matter DE, Matter RM. Suicide among elementary school children: A serious concern for counsellors, Elementary School Guidance and Counselling, 1984; 18: 260–267.
Melear A. Children's conception of death, Journal of Genetic Psychology, 1973; 123: 359–360.
Mishara BL. An empirical investigation of children's understanding of suicide and death. Manuscript in preparation, 1995.
Mishara BL, Tousignant M. Pour une véritable prévention primaire du suicide, Revue Québécoise de Psychologie, 1983; 4(1): 21–31.
Nagy M. The child's theories concerning death, Journal of Genetic Psychology, 1948; 73: 3–27.
National Center for Health Statistics. Vital Statistics of the United-States life Tables, National Center for Health Statistics, Hyattsville, Maryland, 1988.
Normand C, Mishara BL. The development of the concept of suicide in children, Omega International Journal of Death and Dying, 1992; 25(3): 183–203.
Orbach I, Gross Y, Glaubman H, Berman D. Children's perception of death in humans and animals as a function of age, anxiety and cognitive ability, Journal of Child Psychology and Psychiatry, 1985; 26: 453–463.
Pfeffer CR, Lipkins R, Plutchik R, Mizruchi S. Normal Children at Risk for Suicidal Behavior: A two-year follow-up study, Journal of the American Academy of Child and Adolescent Psychiatry, 1988; 27: 34–41.
Piaget J. La Construction du réel chez l'enfant. Neuchatel: Delachaux et Niestlé, 1937.
Raimbault G. L'enfant et la mort. Des enfants malades parlent de la mort: Problèmes de la clinique du deuil. Toulouse: Privat, 1975.
Shaffer D, Fisher P. The epidemiology of suicide in children and young adolescents, Journal of the American Academy of Child Psychiatry, 1981; 20: 545–565.

13. What do we Know about Media Effects on Imitation of Suicidal Behaviour

State of the Art

ARMIN SCHMIDTKE and SYLVIA SCHALLER

I. Problem

It has frequently been suggested that suicidal behaviour, too, might be learnt by a process of modelling and that this may be the case especially for children, adolescents and young adults (Kreitman, Smith, & Tan, 1970, Schmidtke, 1988; Steede & Range, 1989; Platt, 1993). Also explanations why some specific methods or places like bridges are used (e.g. the Golden Gate Bridge; Seiden & Spence, 1982, 1983) are not possible without recurring to theories of social transmission of suicidal behaviour.

In this context effects of contagion on suicidal behaviour were often mentioned. However, from a psychological point of view, the expression and the theory of imitation should be preferred. The reason being, the use of this theoretical framework makes it easier to explain some results, even some contradictory findings, with regard to imitative behaviour in the field of suicidology. In contrast to imitation, contagion implies a kind of infectious disease, without possibilities of the "infected" person to be able to act or to choose for themselves.

In the context of the hypotheses of imitation of suicidal behaviour, the effects of the mass media as transmitters of models for imitation have to be discussed. However, the possible influences of mass media have been a particular source of controversy.

Discussion of this topic has mainly been centered on two questions (Schmidtke & Häfner, 1989):

- Does the specific presentation of suicide themes (and changes therein) by the media merely reflect existing attitudes and opinions, or
- Does the reporting of suicidal acts itself and/or the mode of presentation of such acts by media, influence social attitudes towards such behaviour and in consequence also the suicidal behaviour itself and may it trigger suicidal behaviour?

D. De Leo, A. Schmidtke and R.F.W. Diekstra (eds.), Suicide Prevention, 121–137.

II. Findings

A. Historical Findings

Going back into history, early examples of such modelling can already be found in stories of suicide epidemics in antiquity (e.g. the mass suicide of the Milesian virgins; Singer, 1980, 1984). Here one can already find the preventive effects of "mass media". The rumor, spread by a joker, that the next dead virgin should be carried naked through the market place, immediately stopped the epidemic behaviour. In more recent times, the "Werther" effect is often cited as an example (Phillips, 1974, 1985).

Also in scientific literature, there are early references to suicidal behaviour being caused by imitation (Farr, 1841 (cited in Pell & Watters, 1982); Mathews, 1891 (cited in Phelps, 1911); Strahan, 1893; Phelps, 1911): "Epidemics of suicides or of suicidal behaviour" were reported by Popow, already in 1911. In this context there have also been frequent references to the suggestive power of mass media or similar media like rumors, with respect to suicidal behaviour.

As early as 1910, the historic symposium of the Viennese Psychoanalytical Association in Vienna, dealing with suicides among young students, was already devoted to the question of whether newspaper reports on suicide cause suicides among schoolchildren (Unus Multorum (Oppenheim), 1910).

Also, Phelps suggested already in 1911 that reports on suicides and the treatment of this subject in works of literature could lead to suicidal events induced by imitation. In the same year, a committee of the American Academy of Medicine and the American Medical Association already considered the problem of the modelling effect of press reports on suicide. This committee also quoted examples of increases in the rate of suicide by similar methods (albeit in a rather anecdotal form) when the press had given detailed reports on cases of suicide and the accompanying circumstances (Hemenway, 1911). The influence of newspaper stories was also emphasized by Rost (1912), at this time a famous German journalist and expert in suicidal behaviour.

B. Newer Findings

If we want to evaluate the effects of mass media, we have to classify the behaviour and the method of presentation. With regard to suicidal behaviour, the characteristics of the model behaviour and the model can be divided according to the type of suicidal behaviour, the type of model, and the type of presentation. The model can be real or fictional and the suicidal behaviour can be suicide ideas, a suicide attempt or a suicide. The type of presentation can be in print media (literature), music, broadcasting, movies, TV and stage plays. Figure 1 shows this possible classification.

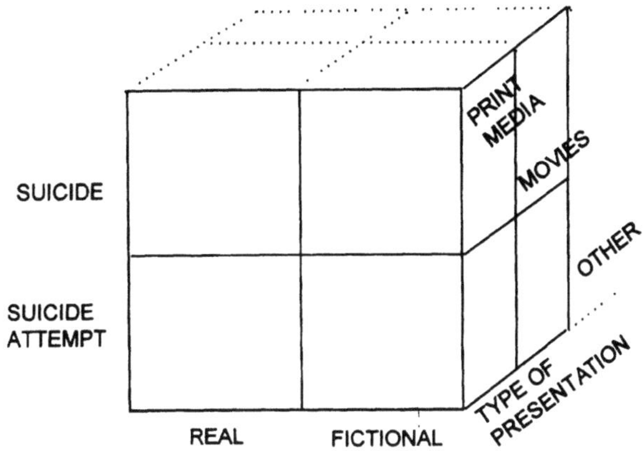

Figure 1. Classification

1. Books: Real and Fictional Suicidal Behaviour

In 1774, Johann Wolfgang von Goethe published his novel "The Sorrows of Young Werther", whose hero finally shot himself after a sentimental, hopeless love-affair.

The story had in its author's own words, "a great, even an immense impact". It launched imitation effects of two aspects: Werther's clothing became gentlemen's fashion of the last quarter of the 18th century and became known as the "Werther dress".

The second possible imitation effect was the widespread impression that Goethe's novel led to an increase of suicides of the same type, i.e. shooting oneself with a pistol. This impression was so strong that authorities in Denmark, Saxony and Milan banned the book (Phillips, 1974, 1985; Voges, 1981; Elsner, 1985).

Therefore, the "Werther effect" has often been cited as the earliest recorded evidence of suicidal imitation effects being produced by literature.

In Hungary, various imitation effects due to two books have also been reported (Fekete & Schmidtke, 1995, 1996). Similar effects have been reported after the publishing of books about suicidal behaviour, especially in connection with books with a content of "how to commit suicide". For example a report by Marzuk et al. (1993) in the New England Journal of Medicine showed an increase of suicides in New York City involving asphyxia by plastic bag, as compared with the numbers in previous years, after the publication of the book "Final Exit".

However, the imitation effects of suicides in literature are difficult to examine, due to time effects – so it is difficult to determine the time of "learning" of the model behaviour – the difficulties of controlling the imitation during a certain period of time etc.

2. Music: Fictional Suicidal Behaviour

The same methodological problems arise when one tries to examine imitation effects of suicidal behaviour in music. A lot of operas contain suicides and suicidal behaviour, and there are also some hypotheses about the imitation effect of popular songs, for example the song "Gloomy Sunday" in Hungary (Fekete & Schmidtke, 1995). There was also a lawsuit brought against the producer of the rock band "Judas Priest", since parents assumed that the song stimulated two young boys to commit suicide. However, due to the problems of time and the difficulties to control concomitant variables it is difficult to take these anecdotical evidences as proof for imitation effects.

3. Stage Plays: Fictional Suicidal Behaviour

In a controlled experiment, Jackson & Potkay (1974) examined the effects on undergraduates of a one-act stage play about suicide, and the possible different effect(s) resulting from its presentation by other media. The suicide potential did not increase.

4. Newspapers: Real Suicidal Behaviour

One of the first studies on the effect of newspapers was the study by Motto. The hypothesis was that during newspaper strikes the absence of models should cause a decrease in suicides of possible imitators. The results of the first study (Motto, 1967) and some replications also by other authors (Motto, 1970; Blumenthal & Bergner, 1973) were contradictory.

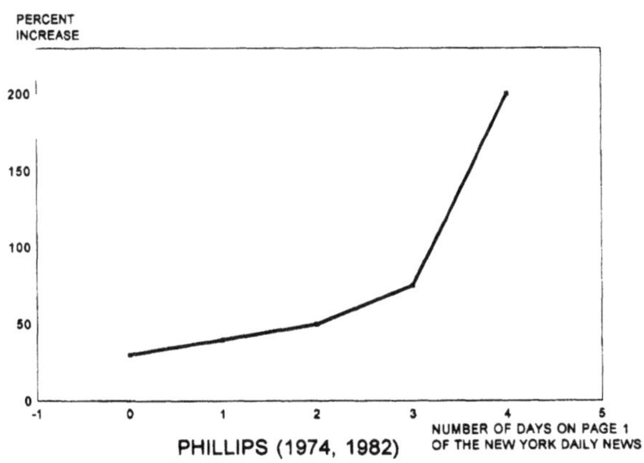

Figure 2. Phillips: Covariation

More convincing are the studies by Phillips on the effects of imitation of suicidal behaviour portrayed in mass media. Phillips found, that the more publicity a suicide case was given – especially on the front pages of newspapers – the more suicides could be found in the period after the publishing of the suicides (Phillips, 1974). This effect could also be found for car accidents Phillips (1977, 1979) and airplane fatalities (Phillips, 1978).

Also some recent studies and case reports show that newspaper reporting of real suicides may lead to imitation (Phillips & Carstensens, 1988; Hassan, 1995; Hills, 1995), however, there are also some contradictory findings like for example the study of Jonas (1992) in Germany. He examined imitation effects for major newspaper reports on suicides of celebrities in a German state and found no significant covariations of the reports with an increase of suicide rates. Stack (1996) investigated the imitation effect in Japan. The increase was similar in magnitude to that reported in the American cultural context, however, the imitation effect was restricted to Japanese victims.

5. Television and Films: Real Suicidal Behaviour

The effects of television reports of real suicidal behaviour are less clear. Bollen and Phillips (1982) found increased suicides up to 10 days after television news reports about suicide. The effects remained significant after correcting for holidays and unequal variability (Phillips & Bollen, 1985). Similar effects were reported by Phillips and Carstensen (1986). The effects of the reporting of suicidal behaviour on accidents were equivocal (Phillips, 1980). However, Marks (1987) criticized the research design used by Phillips and Carstensens as inadequate to test the imitation hypothesis. The crucial test for the imitation hypothesis should be a demonstration that persons who commit suicide are influenced to do so because of their knowledge of a previous suicide (i.e. suicide story). They object that no evidence is presented on this crucial point. Similar objections against this study were published by Mastroianni (1987) (see also Phillips & Carstensens, 1987; Kienhorst, 1994).

Martin (1996) examined the influence of television suicide in a normal adolescent population. 14-year old students claiming more than two exposures to television suicide took more risks and substances, knew more of suicide in real life (the community), had higher depression scores, and had a history of deliberate self harm (DSH) and suicide attempts. In those who reported knowing someone who had died from suicide, frequent exposure to suicide on television appeared to contribute to the variance of suicide attempts. However, it contributed little to either depression or suicidal thoughts. A methodological weakness of this study is that it is not clearly differentiated, whether the television suicides were real or fictional and that when suicide knowledge was controlled for, the relationship between TV suicide and DSH lost significance.

However, in the study by Horton and Stack (1984), the effects were not in favour of the imitation hypothesis. They investigated covariations between the monthly suicide rate and the amount of time (in seconds) given to the coverage of suicide-related stories on the 6 o'clock national (U.S.) news bulletin; unemployment, the divorce rate and seasonal factors were controlled. The results indicated the absence of any relationship between the length of news coverage and the monthly suicide rate.

Up to now, there are no studies examining the effects of the portrayal of a real suicide in films (i.e. films shown in theatres).

6. *Television and Films: Fictional Suicidal Behaviour*

In 1986 Ellis and Walsh reported that, during the week after the broadcasting of an episode of the English soap opera "Eastenders", in which a leading female character took an overdose, an increased number of patients were admitted to their hospital after having taken an overdose. Similar effects were reported in two other English cities (Fowler, 1986; Sandler et al. 1986), whereas no effects were detected in a fourth city (Daniels, 1986). Platt (1987, 1988) has since extended the investigation to 63 English hospitals: although a significant increase in the number of suicide attempts for the total female population compared to the corresponding rates in the year before was detectable, he was unable to find any significant results in the group closest to the "assumed" age to the "model". A further study (Williams et al. 1987) in two accident and emergency departments, also failed to reveal any convincing correlation.

A better example recently quoted as evidence for an imitation effect is the film "Deer Hunter". Since the release of this film in which russian roulette was shown in a scene and in which a leading character of the film later committed suicide applying the same method, 43 young men in the U.S. committed suicide with the same method according to a summary by Coleman (1987). According to the Congress of the American Association for Suicidology in San Diego in 1988 the reported number is now already much higher. This "suicide method" in general is so uncommon, that one can assume imitation effects.

In the context of movie-inspired violence, Wilson and Hunter (1983) reported similar suicides in the U.S.A. after the broadcast of the television series "Death of a Student" (watched in Germany), but they did not provide any statistical comparison. Ostroff et al. (1985) and Ostroff and Boyd (1987) reported (also in a rather anecdotal style) an increase in the number of suicide attempts of adolescents after broadcasting a television movie showing the suicide of a young couple and its effect on their parents.

Phillips (1982) investigated the effect of "suicides" in "soap operas" broadcast on American television networks in 1977. In the second half of the week after such broadcasts, the number of suicides among white Americans (this group was selected because most of the suicide models were whites) significantly increased in comparison to the expected values (the calculation was based on the period before broadcasting and the trend with time). The number of road accidents involving people who were killed or seriously injured (the latter group could only be monitored in California) increased correspondingly. However, in a reanalysis of Phillips' findings, Kessler and Stipp (1984) attempted to correct false assumptions about the time of some of the "suicides" as well as the model behaviour in some cases. They did not reveal any evidence of a linkage between soap-opera suicidal behaviour and subsequent real-life fatalities.

The results of Gould and Shaffer (1986) also provoked a controversial discussion. They examined variations in the number of suicides and suicide attempts in the Greater New York area before and after broadcasting of four TV films dealing with the theme "suicide" in the autumn and winter of 1984 to 1985. During the following two weeks after these broadcasts the mean number of suicide attempts was significantly higher than the mean number observed before. After three of these films there was also a significant increase in the number of suicides when compared with the mean prior to broadcasting.

This study was replicated in some other states in the U.S. However, most of the following studies failed to confirm the first results.

With reservations only, reports dealing with welfare and scientific organizations can be classified as belonging to the group showing fictional suicidal behaviour, as such reports cannot be considered to give information about models in the literal sense. Holding (1974, 1975) for example studied the effects of an eleven-episode television series about the Samaritans ("The Befrienders"), in which each episode showed a predicament leading to a suicidal situation. One episode ended with a suicide. He found a definite increase in the number of contacts with the Samaritans, but no changes in the number of patients admitted to hospital after suicide attempts.

7. Internet: Real and Fictional Suicidal Behaviour

Despite the fact that there is also a growing tendency to deal with suicide themes in new media like the Internet (Fekete, 1996), up to now, no study exists which controls imitation effects (e.g. the effects of instructions of how to commit suicide).

III. Methodological Problems

All previously mentioned findings have been based on "natural observations" or "field experiments", whose validity is limited by the absence of experimental rigour, especially as it has not been possible to rule out competing hypotheses. The first prerequisite for the establishment of a causal relationship between a publicised model and an increase in suicide figures is a clearly defined type of model behaviour, which can also be identified precisely at the level of imitation. Taking basic research findings into account, we suppose that learning by modelling depends on certain characteristics of the model, such as age, sex and social status, and the corresponding characteristics of the observer (Bandura, 1976). The imitation effects also seem to depend on the frequency and scope of presenting the model.

If a significant increase in the modelled behaviour is found, the hypothesis of learning by modelling gains plausibility, provided it is also possible to show that the extent of imitation effects depends on the degree of concordance between certain characteristics of the model and the imitator and the length and scope of presentation. If no allowance is made for age and sex etc., and the type of suicidal behaviour, it is not only impossible to detect the imitation of a specific type of model behaviour, but also the overall modelling effect might elude observation, either because it is too weak or because the increases in certain age (or other) groups are balanced by falling trends in others. Public statistics of causes of death, which were used in the majority of the studies cited above, hardly provide a suitable basis for the assessment of time-related modelling effects: the classification of suicide methods and the determination of the time of death are usually imprecise, and the date and time of a suicidal act is often unknown, at least when "softer" methods of self-harm (e.g. intoxication) are used. Suicide attempts, which account for an essential part of the imitation effect of suicide models, are not included in public statistics and are, in many cases, difficult to assess fully from other sources. To rule out alternative hypotheses, such as seasonal influences or long-term trends, it is also necessary to study enough and sufficiently long control periods.

IV. A Quasi-Experimental Study

In the Federal Republic of Germany, in the 80's and 90's there was a chance to overcome such methodogical problems and to study imitation effects with a natural experiment with replication. The second German TV channel (ZDF) broadcast a six-episode TV serial entitled "Death of a Student" in 1981. The lethal end of the suicidal act (railway suicide) of a 19-year-old male student was repeated at the beginning of each episode, and the beginning of the suicidal act was shown in episodes 2–6. The various episodes of the film showed the events that led to the suicide from the viewpoints of his parents, his fellow students, his teachers, his girlfriend and himself.

The serial was broadcast again in 1982. By using this "natural experiment", we had an opportunity of testing repeated differential imitation effects in suicidal behaviour over a long period of time. The characteristics of the fictional suicide model and the opportunity of obtaining data on its effects enabled to fulfill the most important prerequisites for testing the "Werther effect" hypothesis.

The increase of railway suicides was most marked in those groups whose age and sex closely resembled those of the fictional model (i.e. young men).

For 15- to 29-year-old males (an age grouping comparable to that on which TV audience figures are based was used), 62 suicides occurred for the observation period of 70 days during and after the first broadcasting of the series. The mean value for the other years studied was 33.25, thus indicating an increase of 29 suicides (= 86%).

For females of the same age, 14 suicides occurred, representing a considerably lower increase of 6 suicides (= 75%) as compared to the mean for the other years (8; = 4.07).

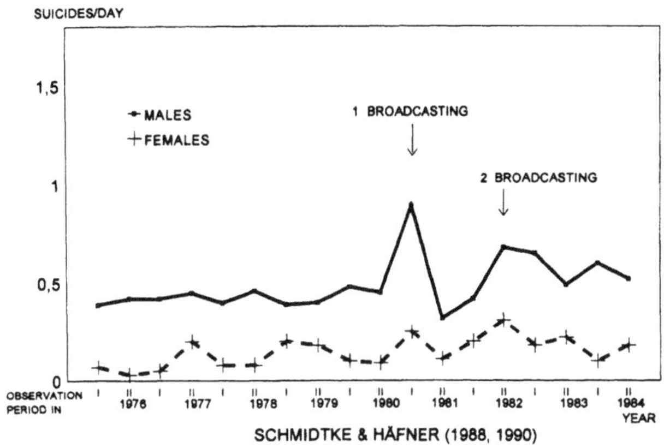

Figure 3. Suicides of 15–29 year old males

A striking finding was also that the frequency of suicides remained higher for a longer period of time after the broadcasting of the model than indicated in Phillips' studies. Suicides in the group of 15–19 year old males remained above the expected number for sixteen weeks. In the group of 20 to 24 year olds, suicides ceased to show any excess after the tenth week.

The effects of the second broadcasting were weaker. In the age group comprising 15- to 29-year-old males, 47 suicides occurred during the period of observation, the increase being approximately 17 (= 54%) in comparison with the mean for the other years investigated (30.38), and approximately 18.5 as compared with the corresponding time periods in the years before and after the broadcasting.

To control influences that might have misleadingly given the impression of there being a correspondence between the variations in the suicide figures and the broadcasting of the suicide model, the various periods were regarded as successive time-related measurements that were standardized by the day on account of the differing lengths of the periods studied. When the two periods of observation and two control periods before the period of broadcasting were considered, a total of 36 units of data were obtained for the whole period. These data were directly comparable and were subjected to time-series analyses. The effect that the film had on 15- to 29-year-old males, i.e. the group closest to the model, was shown most clearly when all the periods were considered simultaneously.

Of all the periods studied – and this was still demonstrable when further control periods were included – the "experimental" period during and after the first broadcasting attained rank 1, while the experimental period of the second broadcasting attained rank 2. The third rank was reached by the first control period, this period starting at the end of the second experimental period. This result is highly significant.

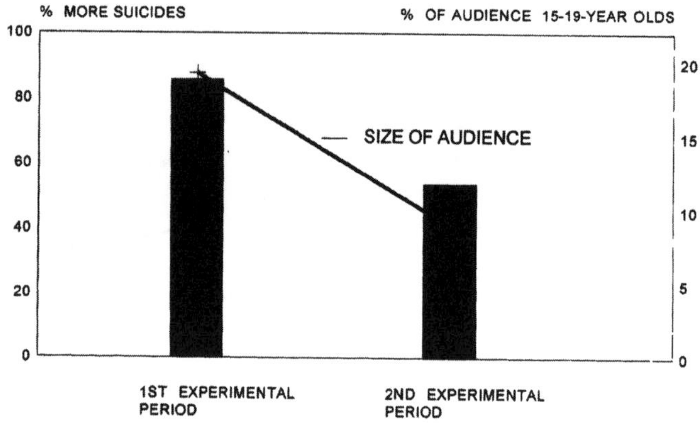

Figure 4. Relations of size of audience and imitation effects

The extent of the increases in the number of railway suicides among 15- to 29-year-old males during and after the first and second broadcasts corresponded to the respective figures of the audience of this age group.

V. Discussion

As frequently pointed out, the results presently available on imitation are riddled with contradictions. However, when the various types of analyses as well as the varying nature of models and behaviour are taken into account, theoretical justifications can be found for these extremely diverse findings.

To study imitation effects it is necessary to understand the conceptual framework of learning by modelling. Learning by modelling refers to the acquisition of new patterns of behaviour through the observation of behaviour of one or more models. Imitation is thus not limited to learning from real models.

The paradigm of modelling has also been extended to include differing ways of perceiving the model. The degree of reality may also vary. The effect of modelling has been experimentally found to depend on the number of models (in the case of media that can mean the number of repetitions), the characteristics of the model (e.g. age, sex, race and social status), the degree to which the behaviour of the model is reinforced, and the charactistics of the observer. Among these characteristics are age (younger people are more suggestible, they imitate their peers more than older persons), gender (this variable influences task specific, females are, for example, more field dependent), self-esteem, self-efficacy (self-efficacy is, according to Bandura (1977), one of the most important variables), extraversion, impulsiveness, mood congruence (that means when model and observer are in the same mood condition) etc. Similarities between specific characteristics of the model and those of the observer play also an important role, even in learning from symbolic models (Groffman, 1982; Groffmann, Kroh-Püschel & Wender, 1982). The influence of race was clearly demonstrated in some studies. Since suicidal behaviour in the US is commonly viewed by African-Americans as a problem for white Americans, African-Americans may be less likely to engage in imitative behaviour, when exposed to white models (Molock, 1994). The study by Stack also showed clearly ethnic "model preferences" (Stack, 1996). According to Meichenbaum, it also seems to be important whether a coping (a model that is showing a coping strategy) or a master model (the model is shown as expert) is available.

Learning by modelling can thus be regarded as a function of certain variables of both the model and the observer. The person variables of the observer may also moderate the probability of exhibiting the learned behaviour. Stimulating events or motivating processes determine whether responses acquired by observation are actually performed at a specific point in time. Therefore, a behavioural strategy learned by observation may be shown even some time after the observation had taken place and the contrary behaviour can even be emitted (Groebel, 1986ab). Therefore, it is also not surprising that extremely publicized mass suicides like the mass suicides of Jonestown, Waco or the mass suicide of the order of the solar temple in Canada, France or Switzerland showed no imitation effects. These results can be explained by the involuntariness of the suicides (e.g. Stack, 1983) and the "nonattractiveness" of some of the group leaders. Also when the models are special celebrities, unknown to a greater audience or special subgroups like youngsters or older people as in the study of Jonas (1992) an imitation effect is unlikely because the model charactericstics are not relevant for possible imitators.

Taking into account the theoretical framework, of the previously published studies dealing with imitation, those of Phillips, in spite of some methodological objections, Ellis and Walsh (1986), Gould and Shaffer (1986) and the studies by Schmidtke and Häfner (1987, 1988) appear to provide firm evidence in favor of the imitation hypothesis for real and fictional suicidal behaviour. These studies have not only demonstrated effects on general suicide rates, but also age- and sex-specific modelling effects with regard to suicidal behaviour, thus confirming the results of basic research. Phillips' studies have revealed covariations in the age of the model and that of the "imitator" (Phillips, 1979) as well as (possible) modelling effects of race and sex (Phillips, 1982).

In the study by Schmidtke and Häfner (1987, 1988), the repeated broadcasting of a clearly defined type of model behaviour involving a specific suicide method and defined model variables along with the opportunity of obtaining complete and precise nationwide data on the resulting modelling effects enabled us to avoid many of the methodological shortcomings of earlier studies and to assess the effect in terms of a "natural" single-case experiment.

The increase in a specific suicide method (e.g. railway suicides) was greatest and longest in groups whose age and sex most closely resembled those of the model.

Therefore, on reviewing the various types of analyses and the different levels of independent and dependent variables dealt with in previous studies, one is led to the general conclusion that the portrayal of suicidal behaviour seems to result in an increase in (or at least a reinforcement of) such behaviour. Multiprogram and highly publicized stories have the greatest impact, more affecting youth or those predisposed (Phillips & Carstensens, 1986, 1988; Phillips, Lesyna & Paight, 1992). Phillips, Lesyna and Paight (1992), therefore, state that media publicity "models" suicide, creating "natural advertisements for suicide".

According to these findings and hypotheses, the World Health Organization lists the influence of media among one of the six most important basic steps for suicide prevention ("toning down reports in the press"; WHO, 1993, p. 39).

In contrast, some studies also show clearly that suicide prevention via media is possible. It is hypothesized that the special media coverage portraying the suicide of the singer Kurt Cobain as an unreasonable act and a useless way of problem solving (especially the interview with his widow) prevented copycat suicides (Berman, Jobes & O'Carroll, 1995).

Also a study by Sonneck and co-workers (Etzersdorfer, Sonneck & Nagel-Kuess, 1992; Sonneck, Etzersdorfer, & Nagel-Kuess, 1993) clearly shows preventive effects of mass media. Since its opening in 1978, the Vienna subway has been used as a method of attempting or committing suicide. The number was very low in early years, but beginning in 1984 there was an increase in suicides and suicide attempts. This trend was neither due to an extension of the Viennesse subway system nor to an increase in the number of passengers. However, the major Austrian newspapers reported these suicides in a very sensational and dramatic way. Therefore, the Austrian Association for Suicide Prevention created media guidelines and requested the press to follow them beginning in June, 1987.

After these guidelines were published, the quality of reporting in general changed markedly. Instead of printing sensational articles, the papers printed either short reports, rarely on the front page, or did not report suicides at all. At the same time, the number of suicides in the subway significantly decreased from the first to the second half of the year 1987, and the rates remained low, as Figure 5 shows.

Therefore, this striking relation between the change in the style of reporting by the print media and the number of subway suicides supports the hypothesis

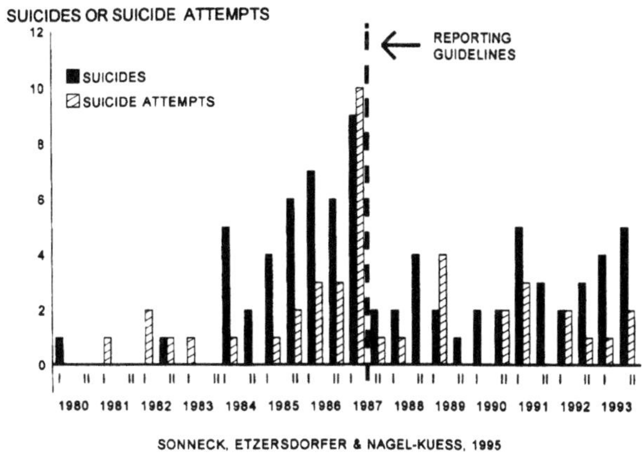

Figure 5. Sonneck study

that reporting and portrayals of real and fictional suicidal behaviour, especially in a certain manner, may trigger additional suicides, and might lead to imitation of suicide methods. As far as we know, the imitation effects of suicidal behaviour depend on the frequency and manner of presenting the model, the model characteristics, the characteristics of the observer, and perhaps also of the method shown. Vice-versa, as the study by Sonneck and co-workers shows, the effects of mass media can also be used for suicide prevention. Another hypothesis that stems from the different findings from Phillips and Schmidtke and Häfner (1987, 1988) is that the reporting of suicidal behaviour and the presentation of fictional suicide models might, theoretically, influence a population's suicidal behaviour in two ways, the short and long term. There may also be different types of responders. "Early" responders may be more impulsive and may have already contemplated suicide; in particular, among young people undergoing a crisis resembling that of the model, the broadcast and reporting may precipitate a suicidal action (e.g. Eisenberg, 1986). In contrast, "late" responders may exhibit a less impulsive, more considered response to a fictional model. The media may so help to "sow the seeds of suicide in the distant future" (Barraclough et al. 1977, p. 531). The broadcasting of suicide-related issues and fictitious suicidal modelling behaviour may lead to the view that this form of behaviour is "a common and understandable way" of problem solving as Littmann already stated in 1985, thus increasing the likelihood of a person choosing suicidal behaviour as a problem-solving strategy in a stress situation. Suicide models may therefore be especially effective in combination with factors such as stress, depression, social isolation, etc., because they might focus on already existing problems, or might induce in the recipient (or viewer) a "morbid rationale" which he or she believes will solve their problems.

References

Altheide DL. Airplane accidents, murder, and the mass media: comment on Phillips. Social Forces, 1981; 60: 593–596.

Bandura A. Lernen am Modell. Klett: Stuttgart, 1976.

Bandura A. Self-efficacy: Towards a unifying theory of behavioural change. Psychological Review, 1977; 84: 191–215.

Baron JN, Reiss PC. Reply to Phillips and Bollen. American Sociological Review, 1985b; 50: 372–376.

Barraclough B, Shepherd D, Jennings C. Do newspaper reports of coroner's inquests incite people to commit suicide? British Journal of Psychiatry, 1977; 131: 528–532.

Berman AL. Interventions in the Media and Entertainment Sectors to Prevent Suicide. Manuscript, 1987a.

Berman AL. Fictional Suicide and Imitation Effects. Manuscript, 1987b.

Berman A, Jobes D, O'Carroll P. The aftermath of Kurt Cobains's suicide. Paper presented at the XVIII International IASP Congress, Venice, 1995.

Blumenthal S, Bergner L. Suicide and newspapers: A replicated study. American Journal of Psychiatry, 1973; 130: 468–471.

Bollen KA, Phillips DP. Suicidal motor vehicle fatalities in Detroit: A replication. American Journal of Sociology, 1981; 87: 404–412.

Bollen KA, Phillips DP. Imitative suicides: a national study of effects of television news stories. American Sociological Review, 1982; 47: 802–809.

Comstock G. Television in America. Beverly Hills: Sage, 1980.

Daniels RG. Emotional crises imitating television (letter). Lancet i, 1986: 856.

Diekstra RFW. A social learning approach to the prediction of suicidal behaviour. In: N Speyer, RFW Diekstra & KJM van de Loo (editors). Proceedings 7th International Conference for Suicide Prevention, Amsterdam, August 27–30, 1973. Amsterdam: Swets & Zeitlinger, 1974: 55–66.

Ellis SJ, Walsh S. Soap may seriously damage your health. Lancet, March, 1986: 686.

Elsner C. Selbstmord/Weil Du mich verlassen hast. Frankfurt: Eichborn, 1985.

Etzersdorfer E, Sonneck G, Nagel-Kuess S. Newspaper reports and suicide. New England Journal of Medicine, 1992; 327: 502–503.

Fekete S. Paper presented at the WPA Satellite Symposium "Suicidal Behaviour". Budapest, 1996.

Fekete S, Schmidtke A. Szuicidium – Modellkövetés – Család. Psychiatria Hungarica, 1995; 10: 131–145.

Fekete S, Schmidtke A. The impact of mass media reports on suicide and attitudes towards self-destruction: Previous studies and some new data from Hungary and Germany. In: B. Mishara (editor). The impact of suicide. New York: Springer, 1996: 142–155.

Fowler BP. Emotional crisis imitating television (letter). Lancet i, 1986: 1036–1037.

Gould MS, Shaffer D. The impact of suicide in television movies. New England Journal of Medicine, 1986; 315: 690–694.

Groebel J. International research on television violence: Synopsis and critique. In: Huesman, LR, Eron, LD (editors). Television and the aggressive child: A cross-national comparison, Hillsdale N.J. Lawrence Erlbaum, 1986a: 259–281.

Groebel J. Fernseh- und Videogewalt: Der aktuelle Forschungsgegenstand. Unterrichtswissenschaft 2, 1986b: 154–167.

Groffmann K-J. Lernen am Modell – Experimente und Theorien. In: G. Seidel (Hrsg.). Orientierungen zum pädagogischen Handeln. Göttingen: Hogrefe, 1982: 51–71.

Groffmann K-J, Kroh-Püschel E, Wender I. A study of social information processing: Some experiments on imitation. In: M. Irle (editor). Studies in decision making: Social, psychological and socioeconomic studies. New York: Springer, 1982: 195–234.

Hassan R. Effects of newspaper stories on the incidence of suicide in Australia: a research note. Australian and New Zealand Journal of Psychiatry, 1995; 29: 480–483.
Hemenway H. To what extent are suicide and other crimes against the person due to suggestion from the press? Bulletin of the American Academy of Medicine, 1911; 12: 253–263.
Hills NF. Newspaper stories and the incidence of suicide. Australian and New Zealand Journal of Psychiatry, 1995; 29: 699.
Holding TA. The B.B.C. "Befrienders" series and its effects. British Journal of Psychiatry, 1974; 124: 470–472.
Holding TA. Suicide and "The Befrienders". British Medical Journal, 1975; 3: 751–752.
Horton H, Stack S. The effect of television on national suicide rates. Journal of Suicide Psychology, 1984; 123: 141–142.
Jackson ED, Potkay CR. Audience reactions to the suicide play Quiet Cries. Journal of Community Psychology, 1974; 2: 16–17.
Kessler RC, Stipp H. The impact of fictional television suicide stories on U.S. fatalities: a replication. American Journal of Sociology, 1984; 90: 151–167.
Kienhorst I. Kurt Cobain. Crisis, 1994; 15: 62–63.
Kreitman N, Smith P, Tan E-S. Attempted suicide as language: An empirical study. British Journal of Psychiatry, 1970; 116: 465–473.
Kreitman N, Smith P, Tan ES. Attempted suicide in social networks. British Journal of Preventive and Social Medicine, 1969; 23: 116–123.
Kreitman N, Smith P, Tan ES. Attempted suicide as language: An empirical study. British Journal of Psychiatry, 1970; 116: 465–473.
Kroth J. Recapitulating Jonestown. Journal of Psychhistory, 1984; 11: 383–393.
Littmann SK. Suicide epidemics and newspaper reporting. Suicide and Life-Threatening Behavior, 1985; 15: 43–50.
Martin G. The influence of television suicide in a normal adolcescent population. Archives of Suicide Research, 1996; 2: 103–117.
Marks A. Television and suicide. New England Journal of Medicine, 1987; 316: 877.
Marzuk PM, Tardiff K, Hirsch CS, Leon AC, Stajic M, Hartwell N, Portera L. Increase in suicide by asphyxation in New York City after the Publication of Final Exit. New England Journal of Medicine, 1993; 329: 1508–1510.
Mastroianni GR. Television and suicide. New England Journal of Medicine, 1987; 316: 877.
Molock SD, Williams S, Lacy M, Kimborough R. Werther effects in a black college sample. Paper presented at the 27th Annual Congress of the American Association of Suicidology. New York, 1994.
Motto JA. Suicide and suggestibility – the role of the press. American Journal of Psychiatry, 1967; 124: 252–256.
Motto JA. Newspaper influence on suicide. Archives of General Psychiatry, 1970; 23: 143–148.
Ostroff RB, Behrends RW, Kinson L, Oliphant J. Adolescent suicides modeled after television movie. American Journal of Psychiatry, 1985; 142: 989.
Ostroff RB, Boyd JH. Television and suicide. New England Journal of Medicine, 1987; 316: 876–977.
Pell B, Watters D. Newspaper policies on suicide stories. Canada's Mental Health, 1982; 30: 8–9.
Phelps E. Neurotic books and newspapers as factors in the mortality of suicide and crime. Journal of Sociological Medicine, 1911; 12: 264–306.
Phillips DP. The influence of suggestion on suicide: Substantative and theoretical implications of the Werther effect. American Sociological Review, 1974; 39: 340–354.
Phillips DP. Motor vehicle fatalities increase just after publicized suicide stories. Science, 1977; 196: 1464–1465.
Phillips DP. Airplane accident fatalities increase just after newspaper stories about murder and suicide. Science, 1978; 201: 748–749.

Phillips DP. Suicide, motor-vehicle fatalities and the mass-media: evidence towards a theory of suggestion. American Journal of Sociology, 1979; 84: 1150–1174.
Phillips DP. Airplane accidents, murder, and the mass media: towards a theory of imitation and suggestion. Social Forces, 1980; 58: 1001-1024.
Phillips DP. The impact of fictional television stories on U.S. adult fatalities: New evidence on the effect of the mass media on violence. American Journal of Sociology, 1982; 87: 1340–1359.
Phillips DP. The Werther efect. Suicide, and other forms of violence, are contagious. Sciences, 1985; 25: 32–39.
Phillips DP, Carstensen LL. Clustering of teenage suicides after television news stories about suicide. New England Journal of Medicine, 1986; 315: 685–689.
Phillips DP, Carstensen LL. Television and suicide. New England Journal of Medicine, 1987; 316: 877–878.
Phillips DP, Carstensen LL. The effect of suicide stories on various demographic groups, 1968-85. Suicide and Life-Threatening Behavior, 1988; 18: 100–114.
Phillips DP, Lesyna MA, Paight DJ. Suicide and the media. In: RW Maris, AL Berman, JT Maltsberger, RI Yufit (editors) Assessment and prediction of suicide. New York: Guilford, 1992.
Platt S. The Aftermath of Angie's Overdose: Is Soap (opera) Damaging to Your Health? British Medical Journal, 1987; 294: 954–957.
Platt S. The consequences of a televised soap opera drug overdose: Is there a mass media imitation effect? In: Diekstra RFW, Maris RW, Platt SD, Schmidtke A, Sonneck G. (editors). Attitudinal factors in suicidal behaviour and its prevention. Amsterdam: Swets & Zeitlinger, 1987.
Platt S. The social transmission of parasuicide: is there a modelling effect? Crisis, 1993; 14: 23–31.
Popow NM. The present epidemic of school suicides in Russia, Newrol. Vestnik, 1911; 18: 312–317.
Rost H. Der Selbstmord in den deutschen Städten. Paderborn: Schöningh, 1912.
Sandler DA, Connell PA, Welsh K. Emotional crises imitating television (letter). Lancet 1986, i: 856.
Schmidtke A. Verhaltenstheoretisches Erklärungsmodell suizidalen Verhaltens. Regensburg: Roderer, 1988.
Schmidtke A. Mass media: Their impact on suicide among adolescents. Paper presented at the International Conference "Suicide: Biopsychosocial Approaches". Athens, 1996.
Schmidtke A, Häfner H. Die Vermittlung von Selbstmordmotivation und Selbstmordhandlung durch fiktive Modelle. Die Folgen der Fernsehserie "Tod eines Schülers". Nervenarzt, 1986; 57: 502–510.
Schmidtke A, Häfner H. The Werther effect after television films – evidence for an old hypothesis. Psychological Medicine, 1988; 18: 665–676.
Schmidtke A, Häfner H. Public attitudes towards and effects of the mass media on suicidal and deliberate self-harm behaviour. In: Diekstra RFW, Maris R, Platt S, Schmidtke A, Sonneck G (editors). Suicide and its prevention – The role of attitude and imitation. Leiden: Brill, 1989: 313–330.
Schmidtke A, Schaller S. The Werther effect: Imitation effects after fictional television suicides. Paper presented at the congress "Psychology and promotion of health". Lausanne, 1991.
Schmidtke A, Schaller S. What do we know about media effects on suicidal behaviour. Paper presented at the XVIII International IASP Congress, Venice. (Abstract Book: 98), 1995.
Schmidtke A, Sonneck G. Introduction. In: Diekstra RFW, Maris R, Platt S, Schmidtke A, Sonneck G (editors). Suicide and its prevention – The role of attitude and imitation. Leiden: Brill, 1989: 163–166.
Seiden RH, Spence M. A tale of two bridges: A comparative suicide incidence on the Golden Gate and San Francisco-Oakland Bay Bridges. Omega, 1983; 14: 201–209.

Seiden RH, Spence MC. A tale of two bridges: Comparative suicide incidence on the Golden Gate and San Francisco-Oakland Bay Bridges. Crisis, 1982; 3: 32–40.
Singer U. Massenselbstmord. Stuttgart: Hippokrates, 1980.
Singer U. Der Massensuizid von Massada bis Guayana. In: Faust V, Wolfersdorf M. (Hrsg.). Suizidgefahr, 1984; 45–55.
Sonneck G, Etzersdorfer E, Nagel-Kuess S. Imitation effect in suicidal behavior: Subway suicide in Vienna. In: K Böhme (editor). Suicidal behavior: The state of the art. Regensburg: Roderer: 66–665.
Stack S. Celebrities and suicide: A taxonomy and analysis, 1948–1983. American Sociological Review, 1987a; 52: 401–412.
Stack S. The Media and Suicide: A Non Additive Model, 1968–1980: A Research Note. Paper presented at the Combined Meeting of the American Association of Suicidology (AAS) and the International Association for Suicide Prevention (IASP), San Francisco, 1987b.
Stack S. Suicide: Media Impacts in War and Peace, 1910–1920: A Research Note. Paper presented at the Combined Meeting of the American Association of Suicidology (AAS) and the International Association for Suicide Prevention (IASP), San Francisco, 1987.
Steede KK, Range LM. Does television induce suicidal contagion with adolescents?. Journal of Community Psychology, 1989; 17: 166–172.
Strahan SA. Suicide and insanity. London: Swan Sonnenschein, 1983.
Unus Multorum[1] Kapitel I. In: Vereinsleitung des Wiener psychoanalytischen Vereins (editor) Über den Selbstmord insbesondere den Schülerselbstmord. Wiesbaden: Bergmann, 1910: 5–18.
Wasserman IM. Imitation and suicide: a reexamination of the Werther effect. American Sociological Review, 1984; 49: 427–436.
Williams JMG, Lawton C, Ellis SJ, Walsh S, Reed J. Copycat Suicide Attempts. The Lancet July 11, 1987; 102–103.
Wilson W, Hunter R. Movie-inspired violence. Psychological Reports, 1983; 53: 435–441.
World Health Organization. Guidelines for the primary prevention of mental, neurological and psychosocial disorders. Geneva: WHO, 1993.

[1] In the English re-edition of the proceedings (Friedman P (editor): On Suicide. With particular reference to suicide among young students. New York International Universities Press, 1967) cited as Oppenheim DE.

14. The Aftermath of Kurt Cobain's Suicide*

ALAN L. BERMAN, DAVID A. JOBES and PATRICK O'CARROLL

Kurt Cobain died at age 27. He was a cultural icon – a hero to a legion of identification-hungry youth. As a celebrity suicide, and as a voice of despair, Cobain's death had all the potential to stimulate vulnerable youth to imitate and follow, to demonstrate the "Werther Effect." This report is a preliminary and site-specific evaluation of that possibility.

Cobain was considered a musical genius. He was a leader in the development of a neo-punk rock music that came to be known as "grunge rock." He was the lead singer/songwriter for the rock band, Nirvana, an extraordinarily successful group formed in 1987.

Two years after the release of their largely unnoticed first album, Nirvana's second release, "Nevermind," went on to sell ten million copies worldwide. Nirvana found superstardom.

Cobain was labeled within the music industry as the next John Lennon. He quickly became the leading symbol of the "lost generation." He pioneered an anti-fashion look (torn jeans; unwashed, chopped and dyed hair; worn out t-shirts; unraveling sweaters), the "grunge uniform." Ironically, grunge, or alternative rock, soon was embraced by mainstream rock stations and fans, and, equally paradoxical, his anti-fashion look soon was being modeled by top fashion designers. Grunge apparel was "in."

Cobain was never able to adjust to his extraordinary success. He was born in 1967 in Aberdeen, a depressed logging town on the coast southwest of Seattle. The older of two children, he was diagnosed as hyperactive as a child and medicated with Ritalin. At the age of eight, his parents divorced. His adolescence found him living with various relatives, an angry, difficult-to-control teen. He had problems in conduct (vandalism) and began what would be a life-long struggle with substance abuse.

In high school, Cobain began playing the guitar. He became increasingly absorbed in heavy metal and punk rock music. Just weeks prior to his high school graduation, he dropped out of school. He did drugs and engaged in petty crime. Two years later Nirvana was born.

After the phenomenal *success of* "Nevermind," Nirvana's third album, "In Utero," followed and also sold millions. Cobain soon met and married Courtney Love (leader of the punk band, "Hole") and had a child. The marriage

D. De Leo, A. Schmidtke and R.F.W. Diekstra (eds.), Suicide Prevention, 139–143.

was tumultuous, characterized by passion, intense fighting and reconciliations, and drug abuse. Public knowledge of their heroin use forced a citation as "unfit parents" and they briefly lost custody of their child.

Cobain also was plagued by chronic stomach pain and severe depression. Four months prior to his suicide, Cobain gave a rare interview published in Rolling Stone Magazine. He commented:

> For five years during the time I had my stomach problems...I wanted to kill myself every day. I came close many times.

He spoke further of his fascination with guns and, in particular, of a song cut from the "in Utero" album entitled, "I hate myself and want to die."

On March 4, 1994, while on tour in Rome, Cobain overdosed on champagne and tranquilizers. Clearly a foiled suicide attempt, his overdose was reported publicly to be an accident.

Nirvana cancelled its European tour and Cobain returned to Seattle to recuperate. Less than 2 weeks later, on March 16, he locked himself in a room with his guns. The police were called by his wife to intervene.

On March 28, after an "intervention" by his wife and friends, Cobain reluctantly entered a drug treatment center near Los Angeles. He stayed less than 2 days, then disappeared. He was not seen again until his body was found (on April 8) by an electrician, sprawled on the floor of a garage apartment at his home, dead of a shotgun wound to the head. He had been dead 3 days. He left a suicide note. Toxicological analyses revealed high levels of heroin and valium in his blood.

As news of Cobain's suicide spread across the U.S. and the world, the grief was intense, pervasive, almost palpable. In the States, radio stations played Nirvana's music around the clock; television, particularly MTV, showed concert footage, interviews, and music videos. Grieving fans gathered at Cobain's residence, leaving notes and flowers. News of his death made the front page of major magazines such as *Newsweek* and *People. Rolling Stone* devoted an entire issue to Cobain.

Two days after his body was discovered, a candlelight vigil was held in Seattle. Some 7,000 fans gathered in a downtown park and listened to a tortured tape, recorded by Courtney Love, in which she tearfully read exerpts from Cobain's suicide note, interspersed with rageful cursing at him.

In retrospect, this event may have been a singularly profound intervention.

Our preliminary data regarding the immediate impact of Cobain's suicide is specific to King County, Washington [metropolitan Seattle], the site of his home and death.

We took a 7-week surveillance period following his suicide. In this period, there were 24 suicides, only one of which appeared linked to Cobain. When compared to the same period, adjusted for day of week, a year earlier, there were fewer suicides in 1994 (1993 prevalence = 31). Figure 1 shows an interrupted time series design with equivalent control groups (1993 and 1995) documenting no marked difference in frequency of suicides after Cobain's. This appears to be good and unexpected news.

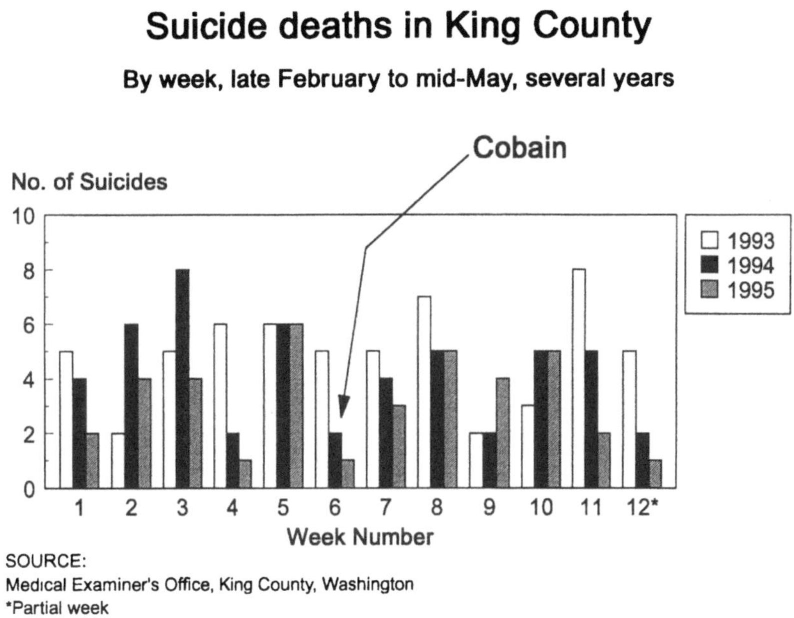

Figure 1.

The one case of suicide clearly linked to Cobain's is a virtual imitation: A 28-year-old male, after attending the candlelight vigil, went home and used a shotgun to shoot himself in the head in the same manner as Cobain. The decedent owned every Nirvana album, was a heavy substance abuser, was isolated and depressed, had previous suicidal thoughts, and had a father who also commited suicide by gun shot wound. His suicide note makes direct and explicit references to Cobain's suicide.

Although this is the only suicide in King County that apparently was influenced by Cobain's, we have learned of others, outside of Seattle, which appear related. For example, in September, 1994, an 18-year-old Canadian youth, who clearly idolized Cobain, shot himself in the head with a shotgun. He owned every Nirvana recording, as well as Nirvana posters and t-shirts. He played guitar and wrote often about Cobain. One poem, entitled "Mad Man," reads as follows:

Jesus Christ commited suicide.
Acting like a wierdo.
I cried when Cobain died.
I lost myself and hero.

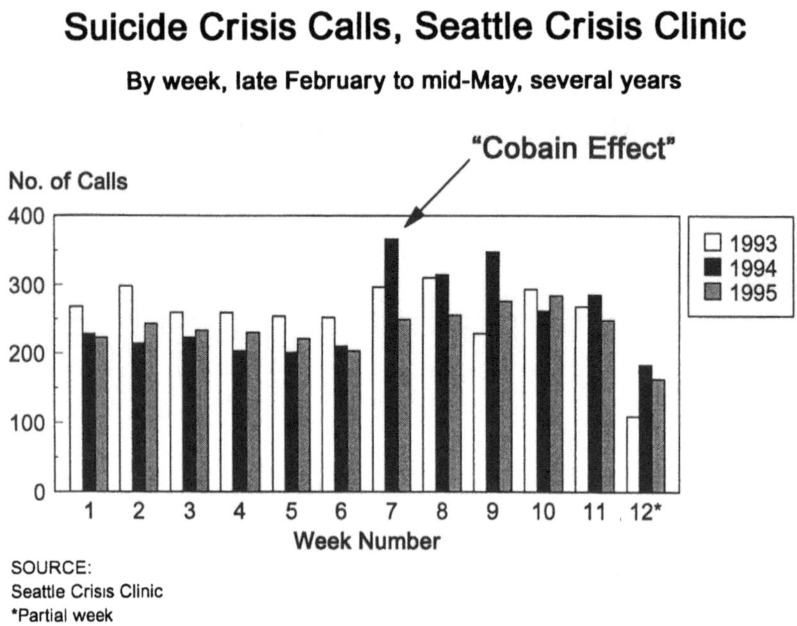

Figure 2.

Three days after Cobain's suicide there was a suicide in San Diego of a 15-year-old female. Her suicide note explicitly stated in closing, "...Remember, I didn't do this because Kurt Cobain shot himself. I had no future with all the unhappiness I experienced and all the disappointments from my parents."

In yet another, perhaps equivocal, case, a 20-year-old Californian man shot himself with a shotgun in front of friends when he propped the gun up to his chin and said, "...Look, I'm just like Kurt Cobain" prior to the gun discharging.

There, no doubt are others. [For example, we have since learned of a 16-year-old suicide in Slovenia; a musician and lead guitarist of his own band, who shot himself after listening to a Cobain CD.] These anecdotal cases describe the expected impact of a celebrity suicide such as Cobain's; but, until more national data is available, we cannot document a statistically significant increase in suicides or suicide rates. Moreover, there may actually be some evidence for an even more positive finding.

As can be seen in Figure 2, the Seattle Crisis Clinic had a clear and demonstrable rise in crisis calls immediately following Cobain's death. These data implicitly suggest the potentially preventive role of sound crisis center work.

I. Conclusions

What can we conclude from the immediate aftermath of Cobain's suicide?

First, much was done right by the media. With few exceptions, the media did well in reporting on Cobain's suicide and life. There was a high degree of professionalism and responsibility exercised in both the print and visual media. A concerted effort was made to distinguish Cobain – the musician – from – the drug abuser – and – the suicide. The general message was: "Great artist, great music, stupid act. Don't do it; here's where to call for help."

Second, Cobain's use of a shotgun countered any romanticized visual image of a lonely, misunderstood star (e.g. , Marilyn Monroe) drifting off into a sleepy, overdose death.

Third, and most importantly, the memorial service that was held after his death was widely covered by the media. It was both tortured and healing. Hearing the grief-stricken taped message of Courtney Love cursing and mourning her lost husband made the death disturbing and real; her honest and open grieving served to deromanticize the death and made it seem profoundly tragic, selfish, and ultimately wasteful. Moreover, the director of the Seattle Crisis Center made a concluding speech at this service. The Crisis Center's number was widely publicized and consistently made available. The result, as evidenced, is that it was used.

Cobain's suicide fits the model of younger, completed suicides, as developed in recent psychological autopsy studies: He had comorbid diagnoses of depression and substance abuse; he owned several guns; he was not compliant with attempted treatment interventions. His suicide forces us to wonder how we can better reach and intervene successfully in the lives of these suicides-about-to-happen. His death continues to describe our failures.

But in the aftermath of this senseless death, there may be only a few imitators. We should be saddened by each. But Cobain's suicide and those of his copycats may have significantly and positively increased public awareness about suicide, crisis services, and available treatment that few other of our efforts could so well accomplish. This, possibly, is the good news that results from tragedy.

Note

*Adapted from Jobes, D.A., Berman, A.L., O'Carroll, P.W., Eastgard, S., & Knickmeyer, S. (1996). "The Kurt Cobain Suicide Crisis: Perspectives from Research, Public Health, and the News Media". Suicide and Life-Threatening Behavior 26(3), 260–271.

15. Media Reports on Suicide in Hungary, Austria, Germany and Lithuania in 1981 and 1991

Reflection, Mediation and Changes of Sociocultural Attitudes Towards Suicide in the Mass Media

S. FEKETE, A. SCHMIDTKE, E. ETZERSDORFER and D. GAILIENE

There is world-wide interest in the effects which the mass media have on human behavior by informing, forming values and attitudes, or influencing forcefully. At the same time the reflection of general social attitudes can be observed and analyzed and cultural valuations and qualifying appear in the way how the media present different human or social phenomena – such as suicide. Empirical studies regarding the relation of suicide and mass media have been conducted in only a few countries and even fewer studies are available concerning the content analysis of the suicidal reports [1, 2, 3, 4]. The suicide models mediated by the mass media can play a role in the sociocultural transmission of suicide; they are part of the suicide culture of a country. There is a remarkable difference between the Hungarian and German or the Austrian and Lithuanian population regarding the ratio of suicide. Analyzing and comparing newspaper reports from these countries seem to reveal some typical cultural variances of the attitudes concerning suicide in the media, which may also play a role in the maintenance of the suicide rate differences. Our aim is to understand these transcultural differences and their changes over time by content analysis of reports on suicide in newspapers from different periods of several countries. We compared headlines about suicide of the West and East German, Hungarian, Austrian, Lithuanian and Greek press from a transcultural point of view.

Material and Method

In our research reports, articles about suicide from daily newspapers – three central and regional papers in each country – were gathered. Data collecting was complete in 1981 and 1991 in 5 countries, the only exception was Greece.

D. De Leo, A. Schmidtke and R.F.W. Diekstra (eds.), Suicide Prevention, 145–156.

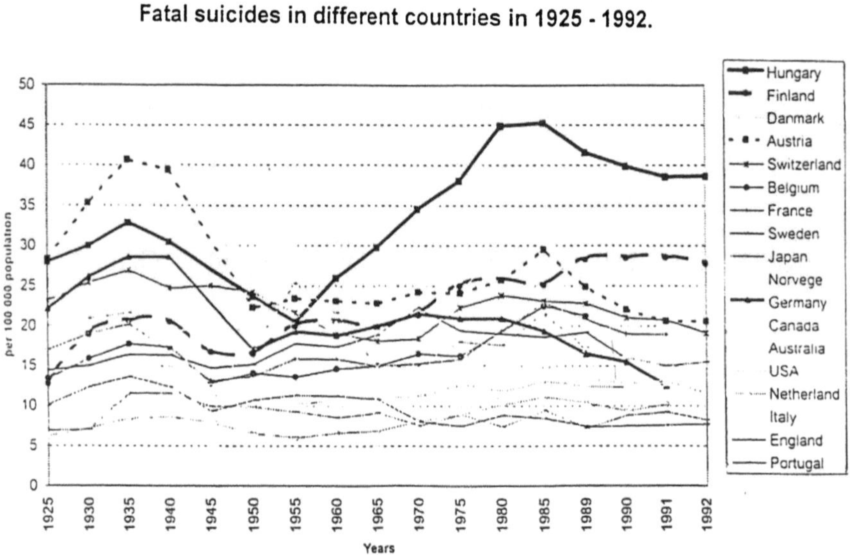

Figure 1.

SUICIDE REPORTS IN 1991

FORMER WEST GERMANY	96
FORMER EAST GERMANY	55
HUNGARY	184
AUSTRIA	143
LITHUANIA	12

SUICIDE REPORTS IN 1981

FORMER WEST GERMANY	209
FORMER EAST GERMANY	5
HUNGARY	60
AUSTRIA	262
LITHUANIA	-

Figure 2.

THE CONTENT ANALYSIS WAS PERFORMED ON THE BASIS OF THE FOLLOWING VARIABLES

Suicide, attempted suicide
Explicit versus ambiguous stories
Prominence of the person
Mentioning of the name, profession, positive or negative characteristics of the prominent person
Suicidal method
Age, sex, motives of the suicidal person
The geographic location
Positive or negative consequences of the suicide
Labeling, qualification
psychiatrization, criminalization,moralization, belittlement
rational judgement, political protest, tragedy, sensation
Murder - suicide
Extended suicide
Prevention, therapy, possible alternatives to the suicide
Statistics, scientific report
Expression referring to the suicidal act

Figure 3.

The samples were compared from the point of view of differences of cultural and social-political systems in the given period. We separated the headlines, subtitles and lines printed in italics, which seem to be the most essential from the view-point of the effect. Readers mostly scan the titles instead of reading the whole story. The title's basic role is to attract attention, to summarize the basic message of the text; it is a trigger. By comparison of the data from the several countries, similarities and differences in the presentation of suicide stories were investigated and interpreted. Our aim was to investigate the changes of public attitudes reflected in the evaluation of suicide events in newspaper reports in 1981 and 1991 as well. Content analyses were carried out on the basis of the occurrence of the following variables in the headlines: labeling, qualification of the suicide, the motives, mentioned positive or negative consequences, prominence of the model person, possible alternatives, expressions referring to the act.

The coding process and statistical analysis were made using the dBASE program for data processing and the SPSS statistical program package.

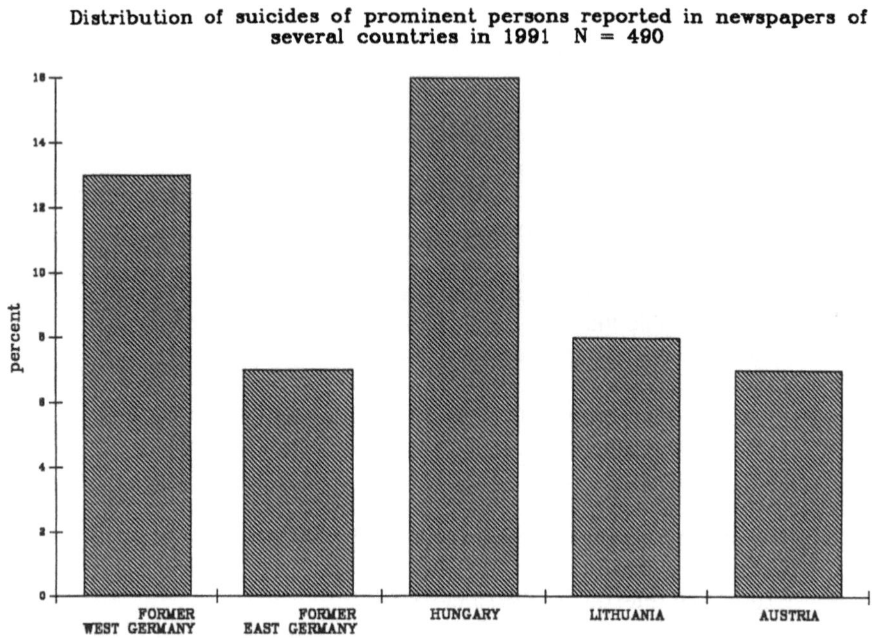

Figure 4.

Results

We collected 1126 headlines of suicide stories in the given periods in these countries. After the coding process a Chi-Square analysis was conducted to see whether any significant differences existed in the rate of the various characteristics in the countries. From the point of view of possible imitation and of the cultural differences the most important findings are demonstrated in the following figures.

The rate of suicide of *prominent* people was similar in West Germany and Hungary and lower in the other countries, but suicides of *negative* celebrities were found significantly more frequent in German texts than in Austrian and Hungarian ones in 1991. The rate of *positive* prominents was the highest in the Hungarian media in 1981 and – to a lower degree – also in 1991. There weren't any significant differences regarding the age, sex of the person and the concrete method of the suicide. The ratio of spectacular methods described in the headlines was higher in each text than in real life. Possibilities for prevention, alternatives to the act rate low in each country. In Figures 7 and 8 the mentioned *negative consequences* of suicide are described as significantly more frequent, and the *positive consequences* as less frequent in German and Austrian headlines than in the Hungarian ones in 1991. The positive consequences are also very frequently demonstrated in the Lithuanian

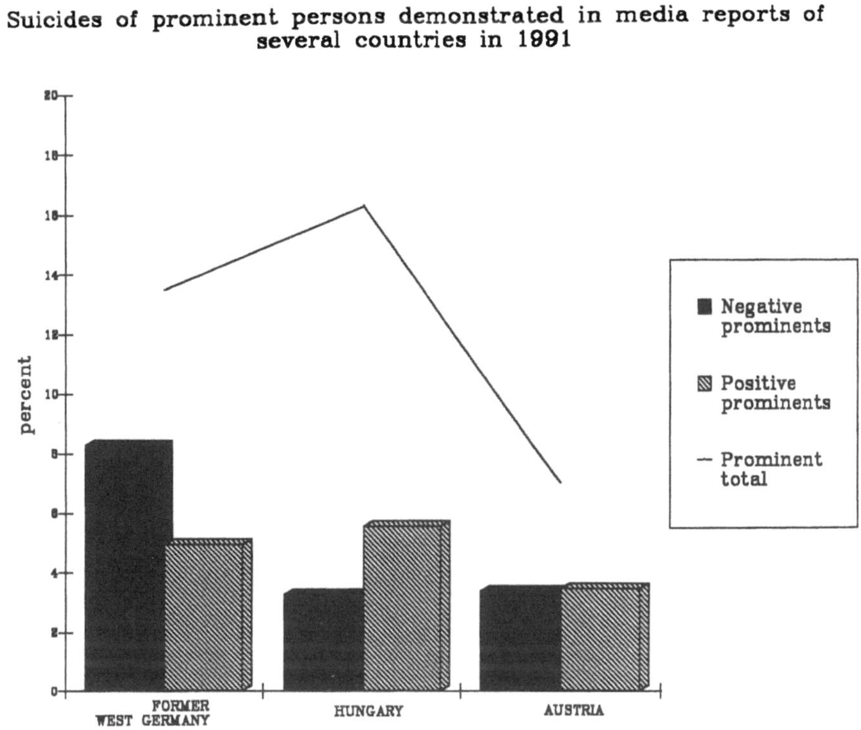

Figure 5.

reports. In 1981 the differences were basically similar, the Austrian data were found between the German and Hungarian ones. The labeling process in the media seems to be very interesting as well. The negative qualifications such as criminalization, psychiatrization, the news about murder-suicide, and about extended suicide can be found much more often in West- and East German and Austrian headlines in 1991 and 1981. The 'positive' accepting labeling – suicide as a tragedy, as a political protest, occurs more often in the Hungarian ones. The labeling of suicide was quite similar in the Austrian and German reports, the only exceptions were protest and tragedy which can be found more frequently in Austrian headlines. It is interesting that the labeling shows more 'tragedy' and at the same time much more 'psychiatrization' in Greek reports than in Hungarian reports. In the Lithuanian material moralization, tragedy and sensational presentation can be observed very frequently. The rate of ambiguous headline stories was higher in the East German, Hungarian and especially in the Lithuanian material (countries with high suicide rate) than in those of the other countries.

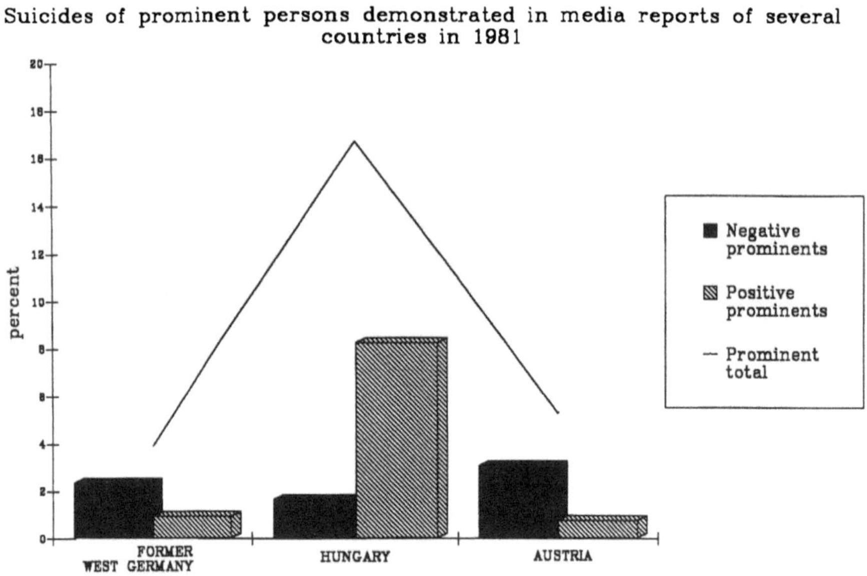

Figure 6.

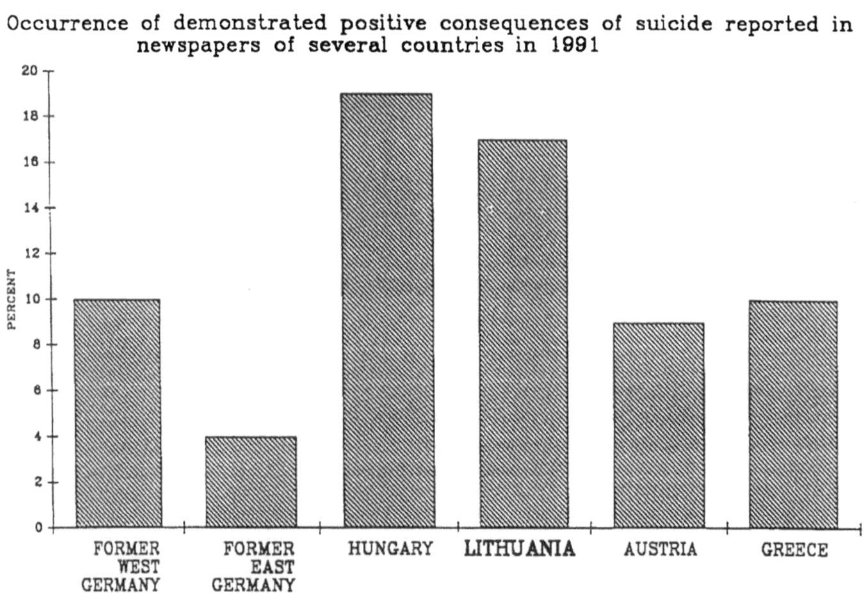

Figure 7.

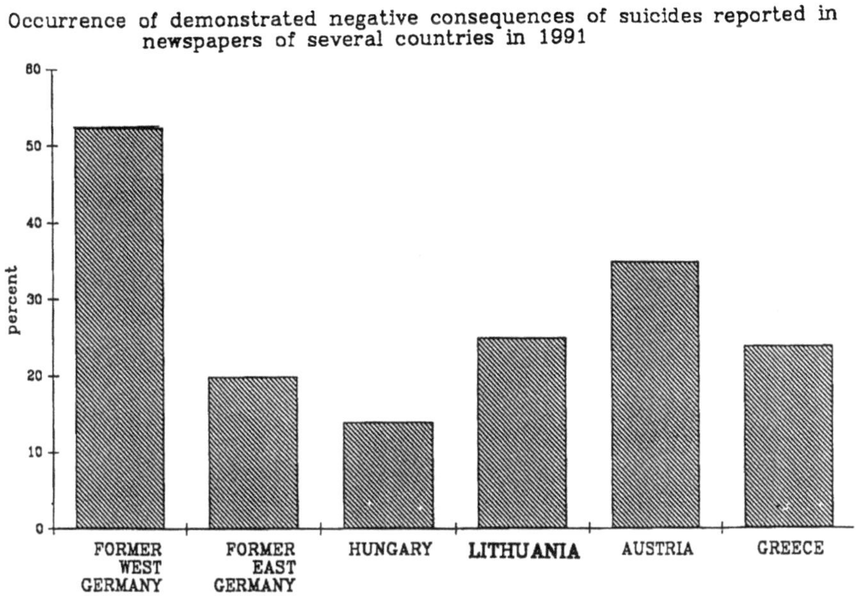

Figure 8.

Discussion

Knowing about the differences between suicidal statistics in the investigated countries, and the fact that suicide is more frequent in Hungary than in the other countries (the only exception is Lithuania and partly GDR) we had the hypothesis that this difference might be observed in the mass media in the differences between cultural valuations and attitudes toward suicide. This sociocultural difference probably contributes to the maintenance of the differences between suicidal rates. In spite of the fact that the suicide rate in the former East-Germany is nearly as high as the Hungarian one, we found only 5 reports on this topic in 1981. The suicide rate is also high in Lithuania, however, there were no reports on suicide in 1981. Beside the known political-social background factors in both countries the presence and attitude-forming effect of West-German electronic media seems perhaps also to be relevant in the former GDR.

Intercultural differences appear in our survey data concerning imitation of suicides of attractive, *prominent* personalities and it's important that the rate of 'negative' prominent people as models was significantly lower in the Hungarian list of patterns, so the identification possibility is higher. It is also significant that in the Hungarian material essentially more positive consequences are shown. By this we mean the 'effectiveness' of suicide, the effects influencing human relationships which evoke positive evaluations

COMPARISONS OF GERMAN, AUSTRIAN AND HUNGARIAN DATA: SIGNIFICANT DIFFERENCES

HUNGARY - AUSTRIA 1991

Significantly more positive consequences are presented in the Hungarian media

Significantly more negative consequences are presented in the Austrian media

Significantly more prominents in the Hungarian media

Significantly more murder-suicides, extended suicides and criminal labeling of the act are presented in the Austrian media

1981

as in 1991, the only exception: differences of positive and negative consequences are not significant

HUNGARY - GERMANY 1991

Significantly more positive consequences are presented in the Hungarian media

Significantly more negative consequences are presented in the German media

Significantly more "positive" labeling /tragedy,protest/ and more sensation in the Hungarian media

Significantly more "negative" labeling /criminalization, psychiatrization are in the German material

Significantly more murder-suicide and extended suicide are in German media

Significantly more "negative" prominents are presented in the German media

More ambiguous headlines can be found in the Hungarian material

1981

the same significant differences can be found

Figure 9.

CHANGES OF ATTITUDES TOWARDS SUICIDE IN THE MASS MEDIA IN GERMANY AND AUSTRIA COMPARING THE YEARS 1981 AND 1991

WEST GERMANY - changes are not significant

Labeling - rate of rational suicides increased
- rate of criminal suicides increased

Rate of prominents, frequency of "negative" and "positive" prominents increased

AUSTRIA - changes are not significant, the only exception:
the rate of presented negative consequences of suicides increased significantly

Labeling - rate of rational suicides increased
rate of criminal suicides increased
rate tragedy-qualification increased
rate of sensational presentations decreased

Rate of "positive" prominents decreased

Figure 10.

in the mass media, showing it as a 'communicative force' or presentations suggesting certain reversibility.

Showing the *negative* consequences of suicide was of a significantly higher rate in the German and Austrian material and this decreases the possibility

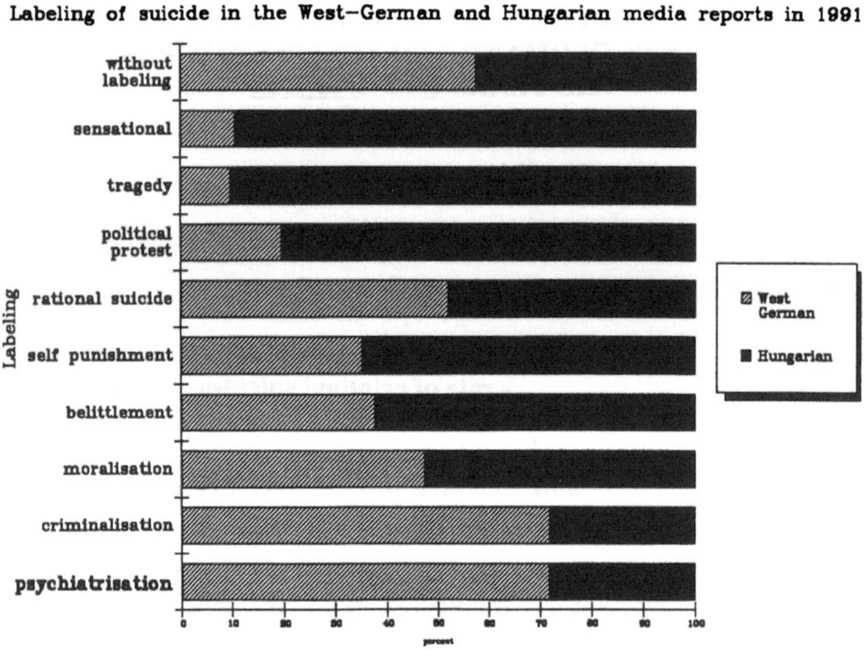

Figure 11.

of imitation-identification. By this we mean the empathic presentation of pains and awful circumstances of suicide, highlighting extended suicide, the joining of murder and suicide, and the consequent sufferings and deaths of other people. The valuations, the *labeling* seen in media presentations reflecting the different attitudes of several cultures toward suicide can either evoke or hinder the model effects of suicide. Negative, onesided valuations of suicide, its criminalization and psychiatrization are much more frequent in the German and Austrian material, while positive, sometimes heroizing evaluation – suicide as a tragedy, as a heroic mode of political protest, are found more often in the Hungarian press. The data from the former East-German media fall in this respect between the representational sample of Hungary and West Germany. It is interesting that in the Hungarian – part of the so-called suicide culture – and in the Lithuanian material the rate of texts which allude only indirectly, implicitly, not evidently to suicide, is much higher than in the German, Austrian or Greek material. The same transfer suggesting uncertainty can be seen in the expressions describing suicide, that is, the rate of wording through the method of indirect speech is much more frequent in Hungarian media. As if an unspoken, obvious sign system based on a consensus, served for the hidden message – "at his early age, due to tragical circumstances deceased unexpectedly". Summarizing, it seems that

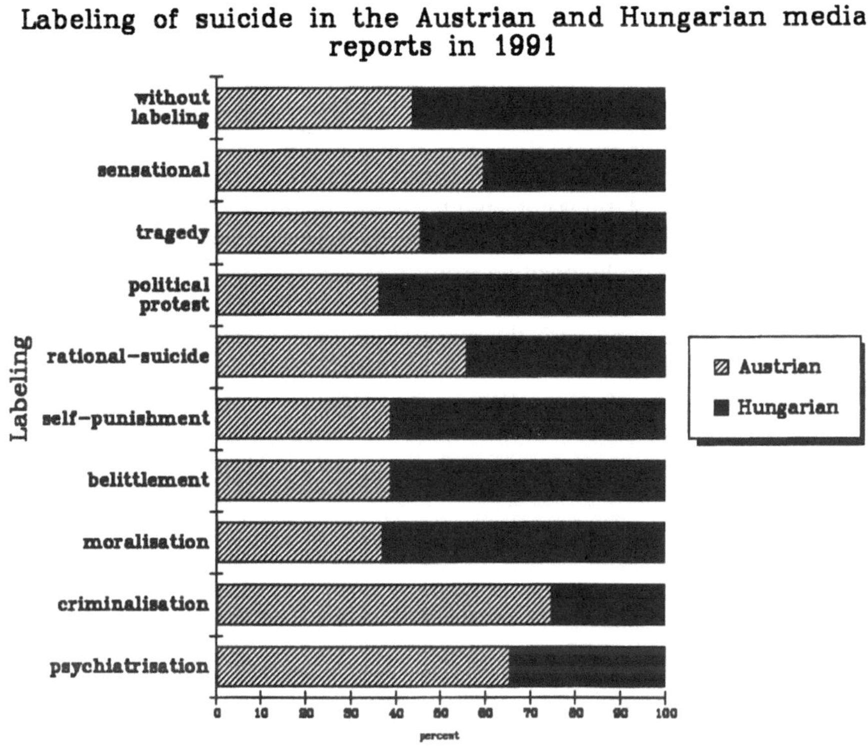

Figure 12.

on the basis of presented consequences, qualifications and wordings in the Hungarian texts, the attitudes toward suicide are rather accepting; the mode of presentation makes imitation rather more possible than do the mass media of other countries [5]. At the same time the high rate of indirect speech, obscure, not evident wordings and euphemisms in the Hungarian mass media may show tendencies of hiding a basically accepting attitude, and may allude to the concerning ambivalence.

Comparing data from the two eras (1981 vs 1991) it can be seen, that in Hungary the spectacular and heroic character of presentations, the moral evaluations as well as the extreme formulations have become less frequent. In contrast the psychiatric and criminal labeling, the openness and the directness of formulations, the preventive and alternative possibilities mentioned, as well as the attempts to understand the suicide act better, occur more frequently. The changes in presentations of suicides in German and Austrian media were not significant. This may be due to the fundamental social and cultural stability in these countries and is consistent with symbolic interaction theory as it has been applied to the effects of the mass media [6].

The modifications in the public attitude concerning suicide reflected in the mass media occurred in a period of great changes in the Hungarian political situation, social structure, and value system. The influence of transition can be considered controversial (downswing in the economy, rapid increase of unemployment, but – in contrast – new community ties and perspectives, strengthening of religiosity, more openness, freedom in society). Our results regarding the way of thinking about suicide in press seem to indicate a general changing trend in this period. On the basis of our content analysis, these changes seem to be positive from the point of view of social learning theory of suicide imitation [7] because of potential decreasing model-effects. Our findings correspond with the cultural and social changes and covaries with the decrease of the suicide rate in Hungary starting in 1988 (compared with continuous increase since 1956). The media are only one feature of the social environment in which suicide behavior can be learned, however, they reflect the essential attitudes and the changes in the culture toward suicide.

References

1. Bandura A. Social foundation of thought and action: a social cognitive theory. Prentice Hall, Englewood Cliffs, 1986.
2. Berman, AL. Fictional depiction of suicide in television films and imitation effects. American Journal of Psychiatry, 1988; 145; 982–986.
3. Fekete, S and Macsai, E. Hungarian suicidal models-past and present. In: Suicidal behaviour and risk factors. 1990, 149–156, Monduzzi Editore, Bologna.
4. Phillips, D. Suicide and the media: research and policy implication. In: Preventive strategies on suicide: A World Health Organization state of the art publication. R.F.W. Diekstra (Ed.), 1990, 1–26.
5. Schmidtke, A, Hatner, H. Public attitudes towards and effects of the mass media on suicidal and deliberate self-harm behavior. In: Suicide and its prevention. 1989, 313–328. R.F.W. Diekstra, R. Maris, S. Platt, A. Schmidtke & G. Sonneck (Eds.), Brill, Leiden.
6. Sonneck, G. et al. Imitation effect in suicidal behavior: subway suicide in Vienna. In: Böhme, K. et al. (Eds.), Suicidal Behavior. The state of the art. Roderer, Regensburg. 1993, 660–665.
7. Stack, S. The effect of the media on suicide: The great depression. Suicide and Life-Threating Behaviour. 1992; 22; 255–266.

16. Shame and Guilt in Suicide and Survivors

NORMAN L. FARBEROW

Two of the feelings most commonly reported by suicidal persons and their survivors are shame and guilt. However, they are not always recognized as such, for the feelings take many forms and are often expressed in different terms, such as self-blame, depression, humiliation, rejection, abandonment, loss, worthlessness, failure, embarrassment, unlovability, and others. Literature search over the past half century indicates more interest in shame than in guilt, perhaps because shame has been seen as the more complex subject. Shame especially may be one of the most basic affects in the area of suicide, so it is of interest that one searches almost in vain for the topic of shame in clinical or theoretical presentations in conferences on suicide and crisis intervention.

Actually it is only from about the 50's on that mental health professionals have explored in depth the role of shame in individual and social development. Earlier, Freud [1] had paid scant attention to shame, referring to it neither as an emotion nor as a symptom but rather as a state of tension, originating within the superego, that developed as the child was forced to control his early tendencies toward narcissistic exhibitionism.

Piers [2], an early analytic writer on shame, accepted Freud's concept of shame as a result of tension, but also expanded it by differentiating the superego into at least two parts, the ego-ideal, or the desirable identity toward which the individual strove as he grew, and the conscience, representing morality, or knowing right and wrong and recognizing acceptable and unacceptable behavior. Such limitations, he said, taught the individual the rules by which one functioned in society.

Anna Freud [3] and Jacobson [4] were among the early psychoanalytic writers on shame and guilt, most of whom focused on its major function of blocking tendencies toward forbidden exhibitionistic behaviors which were considered undesirable, aggressive, narcissistic and immoral. Anna Freud [3] makes the interesting observation, "The qualities of shame, disgust and pity are known not to be acquired by any child except as the result of internal struggles with exhibitionism, messing and cruelty." (in Miller, 5: 157.)

Some writers felt there were other than negative aspects in the concept of shame. Lynd [6], for example, broadened the context for the development of shame by emphasizing the cultural influences on developing affects; shame

D. De Leo, A. Schmidtke and R.F.W. Diekstra (eds.), Suicide Prevention, 157–161.

was seen as the process that facilitates identification with cultural values. Alexander [7] had noted the positive aspects of shame several decades earlier saying that it was the development of feelings of shame over exhibitionism that resulted eventually in a desirable social self. Grinker [8], too, saw shame as being used to facilitate mastery of developmental tasks at the normally expected time.

Theorists in the 60's and 70's began to add other reasons besides conformity and ego-ideal as the basis for shame. Knapp [9] and Kohut [10] postulated that shame's important function was to block arousal which was more threatening to the ego than it was dangerous to morality. Lewis [11] saw shame both as failure to achieve a personal goal established out of admiration for one's parents and as protection against the loss of boundaries of the self. Schneider [12], like Alexander [6] and Grinker [7], also criticized the early views of psychoanalytic writers on shame that focused on the negative debilitating effects of society's use of shame for regulating the individual and failed to acknowledge its role in the development of a person so that he grew up adapted and functioning in both his individual and social environment.

A number of additional feelings (many of them overlapping) were identified during this period as related aspects of shame. These included embarrassment (the core of which is distress over a state of the self that the person defines as no good or not good enough); feeling "undone" and uncomfortably visible (when a part of oneself has been opened up to public view without one's consent or participation); humiliation (as in disgrace, being forced into a debased position, loss of power, loss of dignity); and self-consciousness (being overly aware of self-in-action, feeling others are watching critically).

Some experts have considered guilt as a subset of feelings emerging from shame (Nathanson, [13, 14]; Wurmser, [15]), while others have considered it an independent affect, with its own set of characteristics. Nathanson [13, 14] and Kaufman [16] have built on Tomkins' [17] affect system to make shame central in human development and interpersonal relations in terms of its significance for self-esteem, identity, conscience, intimacy and a sense of dignity. Miller [5] sees shame and guilt as having many elements in common. Her book, "The Experience of Shame", describes shifts between the two feelings occurring so rapidly that conceptual boundaries between them are confused and difficult to maintain. Differentation is possible by noting the direction of the attention of the patient – shame involves attention directed toward some defect exposed in the self-image; guilt attends to the presence or absence of an immoral action.

In sum, the literature identifies both shame and guilt as highly important in both the individuation and socialization of the development of a person's sense of self and relations to others, ensuring conformity with a clear knowledge of what is immoral, sinful and criminal, and facilitating identification with what are esteemed, desirable personal character traits and social behaviors. These are significant areas for both the suicidal person and for their survivors.

The following vignette from a suicide bereavement group illustrates some of the feelings:

Mrs. B, a divorced, Caucasian mother reported the suicide of her adolescent daughter as the culmination of a long and difficult period that included several psychiatric hospitalizations and a number of suicide attempts before the final lethal one. Mrs. B was very depressed, with much self-blame and self-criticism. She had devoted her life to taking care of her daughter after she had become ill, and the loss of her daughter had deprived her of all reason to live. Mrs. B stated before the first meeting that since her daughter had died she thought constantly about killing herself. Question was raised whether she was appropriate for the group. However, she indicated she would be continuing her current individual treatment while attending the group, she was on antidepressants, and most importantly, she had a very supportive circle of close friends. She was accepted in the group with arrangements to keep in close touch by telephone between meetings. After the second session a follow-up call was answered with the news that Mrs. B had ended her life with a massive overdose of pills.

The reactions of the other members in the next quickly arranged session were very similar to those they had experienced after the suicide of their own loved ones. Everyone, members and leaders alike, shared overwhelming feelings of shock, disbelief and loss, along with recollections of everything that had occurred in the first two meetings. Surprise was voiced by most, for Mrs. B had seemed to be responding to the concerned support and caring offered by the group. Strong feelings of both depression and anger were also voiced that she had chosen to act without accepting the available help.

Two major issues were dealt with in this meeting. First, there were the surges of anger at the leaders who had let this happen. The suicide threatened the members' feelings of security in the group and made them all feel more vulnerable to their already strong feelings of depression. Second, one member in particular felt very guilty on learning of Mrs. B's suicide and needed close attention. This member reported that she and Mrs. B had established a bond almost immediately and that they had been in frequent telephone contact after each session. The member reported that Mrs. B had continually urged her to join her in a suicide pact. Mrs. B had also told the member that the only thing that had prevented her from killing herself up to that point was the concern about what would happen to her pets. In their last contact Mrs. B had reported she had found a couple who would take care of her pets "if anything ever happened to her." The other member said that she had not reported the conversations nor the invitation to any one because she had felt that Mrs. B was simply fantasizing the pact and didn't really mean it. Repeated assurances were needed to convince her that she could not have been expected to appreciate the highly serious potential revealed in the repeated invitations and in the search for care for her pets.

Both shame and guilt were strongly mixed in Mrs. B. While there was no shame expressed about her daughter's mode of death, there was deep-rooted shame over her failure as a mother. She had felt both guilt for her child's mental illness and shame that her mothering had not been good enough to make her child well. The suicide of her child confirmed her strong feelings of inadequacy and removed the one factor in her life which had given her some sense of validity. There were also many feelings of guilt in reference to her earlier marital difficulties which had included much turmoil around her alcoholic, abusive former husband.

For the group leaders, Mrs. B's suicide aroused intense feelings of both shame and guilt. This was the only suicide experienced in more than a dozen years of conducting bereavement groups. What was missed? When Mrs. B was called between sessions, she had assured that she was managing with the group's help. What was not appreciated to its fullest was the degree to which Mrs. B had felt life no longer had any meaning. The death of her daughter had deprived her of the main source of her identity. The evidence after her death indicated that she had seen the group primarily as a way of finding someone whom she might persuade to join her in her suicide pact.

The leaders felt that if they had known of Mrs. B's search for a suicide pact partner and for caretakers for her pets, behavior which signaled emergency, they would have recommended to her therapist that she be hospitalized. However, it was learned that she had refused previous efforts to hospitalize her, each time insisting that she was needed to take care of her pets.

The leaders felt guilt for not having appreciated the depth of her loss and for not having sought her hospitalization. But the feelings of shame and embarrassment were more difficult. The leaders were very experienced and had taught others how to conduct bereavement groups. It took many hours of discussion with each other and with colleagues to come to terms with the feelings of self-blame and failure.

In summary, shame and guilt are constantly reported in the experiences of suicidal persons and their survivors. Most often they are disguised or expressed in different terms. While almost always appearing mingled together it is useful to distinguish between them and to recognize their sources. Guilt stems from a moralizing, prohibiting conscience; demanding conformity; the roots of shame lie within the ego invested in personal- and social-ideal behavior. Guilt usually deals with discrete acts; shame with a quality of character revealed by the actions. Guilt refers to a transgression of code or taboo; shame to the sense of inadequacy in meeting standards. Guilt limits strength; shame covers up feelings of weakness. Guilt protects the integrity of objects; shame preserves the integrity of self.

The differences are important for both increasing understanding of dynamics of suicide and its survivors, and for selecting optimal therapeutic response. The alleviation of guilt can be sought through penance, expiation, or some measure of repayment. Mrs. A, for example, was encouraged when she found time for volunteer work in a local children's hospital. Alleviation of feelings of shame is more difficult and more complex. Then the therapeutic goal becomes one of tolerance for perceived self-inadequacies, forgiveness of self and others, and rebuilding of self-esteem.

Survivors of suicide groups continue to be the most effective approach for finding relief from the feelings of shame and guilt. Sharing their loss with others who have experienced the same kind of loss helps in the tolerance of overwhelming pain and facilitates forgiveness of oneself and acceptance of one's humanity.

References

1. Freud S. Character and anal erotism. In J. Strachey (Ed. and Tran.) The Standard Edition of the Complete Psychological Works of Sigmund Freud. Vol. 9. London: Hogarth Press. (Original work published 1909), 1953.
2. Piers G, Singer MB. Shame and Guilt. Springfield: Thomas, 1953.
3. Freud A. Normality and Pathology in Childhood: Assessments of Development. New York: International Universities Press, 1965.
4. Jacobson E. The Self and the Object World. New York: International Universities Press, 1964.
5. Lynd HM. On Shame and the Search for Identity. New York: Harcourt Brace and Co, 1958.
6. Alexander F. Remarks about the relationship of inferiority feelings to guilt feelings. International Journal of Psychoanalysis, 1938, 19: 41–49.
7. Grinker R. Growth inertia and shame: Therapeutic implications and danger. International Journal of Psychoanalysis, 1955; 36: 242–253.
8. Knapp P. Purging and curbing: an Inquiry into disgust, satiety, and shame. Journal of Nervous and Mental Diseases, 1967; 144: 514–534.
9. Kohut H. The Analysis of the Self. New York: International Universities Press, 1971.
10. Lewis HB. Shame and Guilt in Neurosis. New York: International Universities Press, 1911.
11. Schneider CD. Shame Exposure and Privacy. New York: WW Norton, 1911.
12. Nathanson DL. The Many Faces of Shame. New York: Guilford Press, 1987.
13. Nathanson DL. Shame and Pride: Effect, Sex, and the Birth of the Self. New York: WW Norton & Co, 1992.
14. Wurmser L. The Mask of Shame. Baltimore: Johns Hopkins University Press, 1981.
15. Miller S. The Shame Experience. Hillsdale, New Jersey: Lawrence Erlbaum Associates, Analytic Press, 1985.
16. Kaufman G. Shame, the Power of Caring. Rochester, Vermont: Schenkman Books, 1992.
17. Tomkins SS. Affect, Imagery, Consciousness: The Negative Effects, Vol. 2, Northvale, New York: Springer, 1991.

17. Shame – The Unbearable Legacy of Suicide

ONJA GRAD and ANKA ZAVASNIK

I. Introduction

The majority of the literature on shame begins with the observation that this effect has been ignored, little appreciated, or misunderstood [1]. Shame is one of the most intimate feelings the human can feel.

In Darwin's work *The Expression of the Emotions in Man and Animals* published in 1872, he used blushing, which is the expression of self-conscious shame, as a demarcation criterion between animals and mankind [2].

Shame and guilt, in literature usually connected, are difficult both to define and to clearly differentiate. Tantam [2] defines *shame* as being provoked when a person has failed to live up to his or her personal expectations, while guilt may also occur in such situations, but only to the extent that the person acted in the knowledge that such an outcome may result.

A major psychoanalytic statement on shame and guilt was that of Piers and Singer [3]. They defined guilt as the painful internal tension whenever the emotionally charged barrier of the superego is transgressed, as by id impulses of agression and sexuality. *Shame*, on the other hand, results from tension between the ego and that ill defined substructure of the superego the ego ideal. Shame is manifest when a goal of ego-ideal is not attained and therefore is the result of failure [1].

From this distinction between shame and guilt, we come to the differences observed between the sexes. Lewis [4] demonstrates that field-dependent individuals are more prone to the affect of shame and they tend to be women; while more field-independent persons tend to experience guilt and happen to be men.

Lewis et al. [5] proceeded with their investigations on gender differences and shame and found that even at the very early age of three, girls experience and show (!) more shame than boys do when experiencing failure, while they found no difference between them in experiencing pride.

D. De Leo, A. Schmidtke and R.F.W. Diekstra (eds.), Suicide Prevention, 163–166.

II. Shame in our Survivors Group

The decision on our side for preparing this paper was triggered by our clinical practice with the survivors of suicide often experiencing shame, as a very difficult emotion amongst all others – difficult to recognize and even more difficult to express. We noticed that many survivors often cover shame with guilt, stigma, apathy or lowered self-esteem. Even though this ambiguous feeling provokes a lot of problems in the bereaved, it is quite difficult to get a therapeutic grip on this phenomenon.

It is easier to react in the group when the expression of shame is direct, as shown in our case:

The father whose 35-year-old son first killed his mother/wife and than himself, expresses his feelings: "With what you are and that you are, you remind people around you what happened and you put a burden on their shoulders. Suicide is a family seal that never comes off, the whole family is stigmatized. I always have this awful feeling that people talk about our family, so I go to the graveyard only very early in the morning or late at night."

But it is much more difficult for the therapist to help the survivors when the expressions are quite unclear, such as:

A younger widow with two children describes her first socializing after her husband's suicide, when she could dress up as a clown for the Halloween party.

A widower, who subsequently survived suicides of both his wives, calls the therapist at home to let her know that he has no need for help, but wants the addresses of other survivors, because he wants to help THEM.

The father of a 20-year-old girl, who jumped off a building, reports on the group that he feels "humiliated and inferior to the others; everything that he was trying to build, was destroyed. It would be much better if he never got married, as now nothing is left in his life." (In fact he has a wife and a 16-year old son.)

III. Professionals' Reactions after their Patient's Suicide

As shame is usually very painful for the survivors to talk about when they come for help, especially in the form of the research data, we hoped to get some information on this phenomenon by asking colleagues who treat suicidal patients, how they experienced "their" patient's suicide. It was expected that because of their knowledge and training, they would be more open to verbalize their feelings after this kind of loss.

We sent out a questionnaire that contained 15 yes-no questions and 3 descriptive ones to 87 psychiatrists and clinical psychologists throughout Slovenia. We got back 63 completed questionnaires and one which was unanswered.

Our data show that MEN in our sample after their patient's suicide: worked as usual, spoke to their colleagues, became more cautious, and more than half felt guilty. WOMEN needed consolation, spoke to colleagues, became more cautious, and more than three quarters felt guilty.

We found four differences between the sexes that are statistically significant.

More men keep working as usual after suicide (74%), while only half of the women professionals answered they could continue working.

More women needed consolation after the event (30%:71%).

More women had more doubts about their professional knowledge (68%) than did men (27%).

The most enlightening answers for the problem about which we were inquiring are the ones on shame and guilt. On a direct question if they felt shame after their patient's suicide, 31% of female professionals answered positively and all males denied it, while guilt feelings were expressed by 52% of men and 83% of women. Guilt is obviously easily recognized and expressed by respondents themselves. Even though the absolute numbers are small, they confirm previously described gender differences in shame/guilt phenomenon.

The descriptive answers are slightly different. We can not find any category directly connected to shame, but there are some indirect connections:

Men: anguish, horror, uneasiness, failure, ...

Women: anguish, horror, emptiness, helplessness, ...

What might be the explanation for this? Is shame too "shameful" to be self-recognized and acknowledged, so that even the trained professionals prefer to hide it? Or should we adopt Erikson's view that shame is easily absorbed by guilt, which in fact our respondents benevolently expressed – both in the yes/no questions and as a chosen emotional state after suicide.

If we divide the professionals by the years of work (under ten years and more than ten years), we find three differences: younger colleagues tend to feel significantly more shame than do the more experienced ones, they also need to talk to their friends more and they tend to avoid professional discussions.

IV. Discussion

By reconsidering the phenomenon of shame we wanted to find ways of helping the survivors to recognize their emotions including shame and guilt and verbalize them to their relatives or in therapy, using the answers of the professionals being suicide survivors as well.

Interpreting the results of the questionnaire, more new questions than answers have arised:

- Why is shame more expressed by the younger therapists?
- Why is shame more expressed by women?

- Is shame difficult to recognize?
- If it is recognized, what prevents shame from being verbalized?
- Could shame be hidden in other emotional categories?
- What is the connotation of the word shame?

Suicide obviously provokes a turmoil of different emotions, some of them ambiguous and difficult to differentiate. Shame and guilt are the most unrecognizable and misty of them all. What proportion of each the survivor will recognize and express will depend on the mixture of different variables.

In professionals, gender and years of work proved to be the most important differentiating variables for expressing shame in our sample. Descriptive answers of the questionnaire, expressing anguish, grief and guilt, seem very similar to the ones we get from the bereaved after the suicide of a relative. Thus, the term survivor of suicide should stand not only for the relatives, but for the therapists as well.

As is well known, both clients and professionals need help after suicide occurs. Not only do the survivors need time and an opportunity to talk, but it is also necessary that a consultant with the patient or with a professional pays attention to the subtle signs of vague and mixed feelings that can represent or hide shame.

References

1. Morrison AP, Shame, Ideal Self and Narcissism. Contemporary Psychoanalysis, 1983; Vol.19/2: 295–318.
2. Tantam D. Shame and Groups, Group Analysis, 1990; Vol. 23: 31–43.
3. Piers G, Singer M. Shame and Guilt. Norton: New York, 1953: 212–23.
4. Lewis HB. Shame and Guilt in Neurosis. New York: International University Press, 1971.
5. Lewis M, Alessandri SM, Sullivan MW. Differences in shame and pride as function of children's gender and task difficulty. Child-Dev, 1992; Jun:63(3): 630–638.

18. The Erwin Ringel Memorial Lecture

On Suicide and Mental Illness: How Right was Ringel?

JOUKO LÖNNQVIST

In 1949 Erwin Ringel examined 745 suicide attempts, not suicides, with the special aim, as he said, "of finding out their psychic state prior to their suicidal act". Based on this study Ringel felt justified in speaking of a presuicidal syndrome, consisting of three elements: constriction, inhibited aggression turned against the victim's self, and suicide fantasies [1].

I. Presuicidal Syndrome and Mental Disorders

For Ringel human life meant possibilities for creative action and development. In the presuicidal state these possibilities are replaced by an experience of harassment, of being surrounded from all sides and of being ever more intensively squeezed into a steadily tightening space. Ringel saw constriction as situational, dynamic, constriction of human relations and values.

Dynamic constriction includes such fixed patterns of behaviours which remind of personality disorders and such affective constriction which refers to depressive states. Ringel followed directly the thinking of Freud by stating that aggressive impulses turning against the victim's self constitutes the first decisive stage in the presuicidal syndrome. Endogenous depression was a classical example of inhibited aggression to Ringel. Whereas he did not connect suicidal fantasies to any specific psychiatric disorders.

To Ringel the concept of presuicidal syndrome provided better understanding of the psychopathology of suicide, better diagnosing and detecting potential suicide tendencies, and making available specific "anti-suicide therapies". Ringel stated that the presuicidal syndrome is not a part of any psychiatric disorder but rather constitutes a common denominator of all psychiatric disorders that may result in suicide. He declared that psychiatric examination, diagnosis and effective treatment play a central role in all suicide prevention [2, 3].

D. De Leo, A. Schmidtke and R.F.W. Diekstra (eds.), Suicide Prevention, 167–172.

TABLE 1

Retrospective diagnostic picture of suicides

Diagnosis	Proportion (%)
Affective disorder	26–80
Alcoholism	15–54
Schizophrenia	2–11
Organic disorders	1–4
Drug abuse	1–45
Anxiety disorders	1–9
Adjustment disorders	0–14
Personality disorders	3–9
No diagnosis	0–19

Ringel taught that diagnostically one deals primarily with five patient groups having high suicide risk: endogenous depression, neurotic reaction, neurosis, depression in old age and alcoholism.

In his opinion more than one fourth of all suicides are melancholics, and depressive patients in general have very serious suicide risk [4]. In treating these patients modern antidepressants were needed, and even preventive drug treatment (lithium) was one alternative suggested by him.

In neurosis, or as Ringel said, in "genuine neuroses" suicide very often was to Ringel the logical termination of a chronic distortion of life. Its treatment was much more difficult, and required almost always individual psychotherapy.

Old age depression, and problems connected to this, extended furthest beyond the scope of psychiatry. In Ringel's opinion they could be resolved only by a new responsible ethical attitude towards old people, and by improved social and mental health prevention, combined with appropriate somatic care.

Ringel suggested that all alcoholics should be referred to appropriate special agencies for withdrawal, continuation care, and after-care. In their treatment medical, psychological and social interventions should be used.

II. Contemporary Views on Suicide and Mental Disorders

What do we know today about psychiatric disorders as major risk groups of suicide and how do we assess the position of depression, neurotic disorders, old age problems and alcoholism as specific risk groups of suicide?

We know from epidemiological follow-up studies and from psychological autopsy studies [5–12] that mental disorders are carrying a markedly high-

TABLE 2

DSM-III-R diagnoses in a sample of consecutive suicides (N=229)

Diagnosis	Males (N=172)	Females (N=57)
Any Axis I disorder	92%	96%
Major depression	26%	46%
Other depressive disorder	30%	21%
Alcohol dependence/abuse	48%	26%
Drug dependence/abuse	5%	5%
Schizophrenia	8%	5%
Other psychoses	7%	7%
Anxiety disorders	10%	14%
Adjustment disorder	4%	7%
Other Axis I disorder	7%	4%
Any Axis II disorder	30%	33%
No diagnosis of mental disorder	2%	0%

Multiple diagnoses allowed
Insufficient information for Axis I assessment 5% (N=11)
Insufficient information for Axis II assessment 19% (N=44)

tened risk of suicide (Table 1). The importance of a psychiatric approach to suicide lies in finding out the disorder-specific risk factors and specific interventions for individual disorders. However, the role of mental disorders in suicides is often ignored.

We have conducted a series of studies [13–29] in Finland which show tight connections between suicide and mental disorders. In 93% of suicides the victims received current clinical syndrome on AXIS I (Table 2). The most common disorders were depressive syndromes (66%), which included major depression, depressive disorder not otherwise specified, bipolar disorders, dysthymia, depressive schizoaffective disorder, organic mood disorders, and adjustment disorders with depressive mood. Almost one third of the victims (31%) received the diagnosis of major depression. The prevalence of major depression was higher in woman (46%) than in men (26%). If we take the victims aged 60 years or older, depressive syndromes were diagnosed in 74% of all elderly and 65% of the younger victims. The prevalence of major depression was significantly higher in the elderly.

Three quarters of major depressive victims had a history of psychiatric treatment, but only 45% were receiving psychiatric care at the time of the suicide. Most suicide victims were found to have received no specific treatment for major depression. Only 3% had received antidepressants in doses equivalent to 150mg of tricyclic antidepressants. Weekly psychotherapy by a therapist with relevant training in individual psychotherapy was received by 7%. Electroconvulsive therapy was received by only 3% during the last 3 months before suicide. Overall, if strict criteria are used to define adequate treatment, it seems that almost all suicides in major depression occur in untreated or undertreated cases. However, the comorbidity present in 83% of these cases, is likely to set limits to the effect of improved recognition and treatment of depression in promoting suicide prevention.

Anxiety disorders was found in 11% of the cases; only three from 229 suicides were suffering from current panic disorder. It is likely that, in completed suicides, current panic disorder without comorbidity is quite rare. Unlike the adult suicides, adjustment disorders were common in adolescent suicides, diagnosed in 25% of males but not seen in female adolescent victims.

Alcohol dependence or abuse was found in 43% of the cases. Alcohol dependence was more common in men (39%) than in women (18%). Alcohol abuse was also more common in younger victims than in the elderly. Nevertheless, even in victims aged over 60, alcoholism was clearly the second most important disorder underlying suicide. Although the importance of alcoholism in suicide decreases with increasing age, elderly alcoholism should not be neglected when assessing the risk of suicide among elderly patients. Alcohol abuse or dependence was found less often in the adolescent victims (26%) than in adult suicides. Male alcoholic suicides had experienced separations more commonly (28% vs. 5%) and family discord (38% vs. 7%), financial trouble (29% vs. 7%) and unemployment (23% vs. 2%), whereas the nonalcoholic depressive males had had more somatic illnesses.

In our study schizophrenia and schizoaffective disorder was diagnosed in 10% of the cases. The proportion of cases with psychotic disorder was 25% in total.

Personality disorders among suicide victims have been specified in only a few studies. Our finding, one third (31%) having some personality disorder, is comparable with the results from the previous studies. Our data also suggest that current life events, prior to suicide, are especially common among alcoholics who had comorbid personality disorders.

III. Concluding Remarks

Erwin Ringel was a strong personality. He had opinions and visions. He believed in the concept of the presuicidal syndrome. As a researcher and psychiatrist Ringel saw the many connections between mental disorders and suicide. Today it is easy to say, that his experiences come mainly from suicidal patients and attempted suicides, not as much from completed suicides. In addition, he missed many details, stressed suicide risks of neurotic disorders too much, and did not see the existence and extent of comorbidity. However, in total he listed the main risk factors: depression, substance abuse, personality disorders and psychoses. He also suggested treatment strategies for suicide prevention which could be described as eclectic. As a professional I see Erwin Ringel's thoughts on suicide and mental illness as moderate and integrative. I deeply appreciate his work.

References

1. Ringel E. The pre-suicidal syndrome. Psychiatria Fennica, 1973; 4: 209–211.
2. Ringel E. Grundlagen der modernen Selbstmordverhütung (The bases of the modern suicide prevention). Psychiatria Fennica, 1973; 4: 203–208.
3. Ringel E. Suicide prevention in Vienna. In Resnik HLP (editor). Suicidal Behaviors – Diagnosis and Management. Churchill Ltd, London, 1968.
4. Ringel E. Neue Untersuchungen zum Selbstmordproblem (New investigations on the problem of suicide). Hollinek, Vienna, 1961.
5. Robins E, Gassner S, Kayes J, Wilkinson RH Jr, Murphy GE. The communication of suicidal intent: a study of 134 consecutive cases of successful (completed) suicide. Am J Psychiatry, 1959; 115: 724–733.
6. Dorpat TL, Ripley HS. A study of suicide in the Seattle area. Compr Psychiatry, 1960; 1: 349–359.
7. Barraclough BM, Bunch J, Nelson B, Sainsbury P. A hundred cases of suicide: clinical aspects. Br J Psychiatry, 1974; 125: 355–373.
8. Beskow J. Suicide and mental disorder in Swedish men. Acta Psychiatr Scand, 1979; 277(suppl): 1–138.
9. Chynoweth R, Tonge JL, Armstrong J. Suicide in Brisbane: a retrospective psychosocial study. Aust NZ J Psychiatry, 1980; 14: 37–45.
10. Rich CL, Young D, Fowler RC. San Diego Suicide study, I: young vs old subjects. Arch Gen Psychiatry, 1986; 43: 577–582.
11. Arato M, Demeter E, Rihmer Z, Somogyi E. Retrospective psychiatric assessment of 200 suicides in Budapest. Acta Psychiatr Scand, 1988; 77: 454–456.
12. Åsgård U. A psychiatric study of suicide among urban Swedish women. Acta Psychiatr Scand, 1990; 82: 115–124.
13. Marttunen MJ, Aro HM, Henriksson MM, Lönnqvist JK. Mental disorders in adolescent suicide: DSM-III-R axes I and II among 13 to 19 year olds. Arch Gen Psychiatry, 1991; 48: 834–839.
14. Isometsä ET, Henriksson MM, Lönnqvist JK. Completed suicide and recent lithium treatment. J. Affective Disord, 1992; 26: 101–104.
15. Marttunen MJ, Aro HM, Lönnqvist JK. Precipitant stressors in adolescent suicide. J Am Acad Child Adolesc Psychiatry, 1993; 32: 1178–1183.
16. Henriksson MM, Aro HM, Marttunen MJ, Heikkinen ME, Isometsä ET, Kuoppasalmi KI, Lönnqvist JK. Mental disorders and comorbidity in suicide. Am J Psychiatry, 1993; 150: 935–940.
17. Isometsä ET, Henriksson MM, Aro HM, Heikkinen ME, Kuoppasalmi KI, Lönnqvist JK. Suicide in major depression. Am J Psychiatry, 1994; 151: 530–536.
18. Isometsä ET, Henriksson MM, Aro HM, Lönnqvist JK. Suicide in bipolar disorder in Finland. Am J Psychiatry, 1994; 151: 1020–1024.
19. Isometsä E, Aro H, Henriksson M, Heikkinen M, Lönnqvist J. Suicide in major depression in different treatment settings. J Clin Psychiatry, 1994; 55: 523–527.
20. Isometsä E, Henriksson M, Aro H, Heikkinen M, Kuoppasalmi K, Lönnqvist J. Suicide in psychotic major depression. J Affective Disord, 1994; 31: 187–191.

21. Marttunen MJ, Aro HM, Henriksson MM, Lönnqvist JK. Psychosocial stressors more common in adolescent suicides with alcohol abuse compared with depressive adolescent suicides. J Am Acad Child Adolesc Psychiatry, 1994; 33: 490–497.
22. Heikkinen M, Aro H, Henriksson M, Isometsä E, Sarna S, Kuoppasalmi K, Lönnqvist J. Differences in recent life events between alcoholic and depressive non-alcoholic suicides. Alcohol Clin Exp Res, 1994; 18: 1143–1149.
23. Marttunen M, Henriksson M, Aro H, Heikkinen M, Isometsä E, Lönnqvist J. Suicide among female adolescents. Characteristics and comparison with males in the age group 13 to 22 years. J Am Acad Child Adolesc Psychiatry, 1995; 34: 1297–1307.
24. Isometsä E, Heikkinen M, Marttunen M, Henriksson M, Aro H, Lönnqvist J. The last appointment before suicide: is suicide intent communicated? Am J Psychiatry, 1995; 152: 919–922.
25. Isometsä E, Henriksson M, Marttunen M, Heikkinen M, Aro H, Kuoppasalmi K, Lönnqvist J. Mental disorders in young and middle aged men who commit suicide. BMJ, 1995; 27: 145–146.
26. Isometsä E, Heikkinen M, Henriksson M, Aro H, Lönnqvist J. Recent life events and completed suicide in bipolar affective disorder. A comparison with major depressive suicides. J Affective Disord, 1995; 33: 99–106.
27. Henriksson M, Marttunen M, Isometsä E, Heikkinen M, Aro H, Kuoppasalmi K, Lönnqvist J. Mental disorders in elderly suicides. Int Psychogeriatrics, 1995; 7: 275–286.
28. Heikkinen ME, Isometsä ET, Marttunen MJ, Aro HM, Lönnqvist JL. Social factors in suicide. Br J Psychiatry, 1995; 167: 747–753.
29. Lönnqvist JK, Henriksson MM, Isometsä ET, Marttunen MJ, Heikkinen ME, Aro HM, Kuoppasalmi KI. Mental disorders and suicide prevention. Psychiatr Clin Neurosciences, 1995; 49 (Suppl): 111–116.

19. A Critical Evaluation of Psychotherapy in the Treatment of Depression and in Suicide Prevention

DANUTA WASSERMAN

I. Introduction

In his research and clinical work, Professor Erwin Ringel tried to elucidate an important question: whether any specific antisuicidal psychotherapy exists. His answer was that it does [1]. However, in spite of their psychological needs, most suicidal patients lack motivation for psychotherapeutic treatment. On the contrary, they want to take their own lives and escape from all earthly problems. Ringel states that it is the doctor's duty – according to the Hippocratic oath – to do his utmost to enter into a therapeutic process and dialogue with a suicidal patient. In Ringel's view, there are no contra-indications for psychotherapy of suicidal patients – neither age, intellect, diagnosis, asocial tendencies nor psychopathy. Ringel was influenced by VE Frankl and Homburger Eriksson, and in his work tried to unite the principles of individual psychology with a social life framework.

According to Ringel, the objectives of psychotherapy with suicidal patients are as follows:

- to rapidly overcome the presuicidal syndrome
- to create a positive life framework for the suicidal patient
- to develop the suicidal patient's resources, including power of initiative and independence.

Ringel advocates an attitude of active therapeutic encouragement in the acute phase and an analytical attitude in the long run, and states that it is important to try to understand causes of destructive behaviour and the unconscious processes underlying the suicidal development. The doctor/therapist is entitled in the acute situation to circumvent all strict therapeutic rules in order to try to save the suicidal person's life at all costs. This kind of active therapeutic approach puts heavy pressure on the therapist and can contribute to several negative countertransference reactions on the therapist's part [2].

D. De Leo, A. Schmidtke and R.F.W. Diekstra (eds.), Suicide Prevention, 173–183.

Psychoanalysis and classic psychotherapy are not considered by Ringel to be primary methods of choice for suicidal patients. In certain cases, however, suicidal patients are receptive to psychoanalysis. Moreover, suicide problems are touched on in every psychoanalysis as a result of repressions being eliminated and the repressed sides of the patient's self – those he or she has difficulty in tolerating – being brought to the fore.

Ringel described the following particular features in the treatment of suicidal patients:

1. *An approach characterized by a special bond between patient and therapist.* He rightly thinks that it is important to develop a tenable therapist-patient relationship and, through it, to try to keep the patient alive and not lose him or her. This means that the patient's ambivalence must be overcome and the patient's transference, which may be both positive and negative, understood. Ringel's conclusion is that it is important to be aware of daring to bond the patient to oneself in the acute phase of the treatment.
2. *Defusion of aggressiveness.* The immediate danger of suicidality decreases with verbalisation, which occurs in psychotherapy, and accordingly with the defusion of aggressiveness. In the longer term, however, it is essential to work on the suicidal patient's unconscious aggressive desires for self-chastisement, e.g., arising from unconscious feelings of guilt, and on the unconscious conflicts, making them conscious. Ringel emphasises that it is important to analyse the patient's motives and unconscious fantasies, and only in cases where the patient is unable to achieve insight into his or her problems may the doctor work suggestively, use autogenic training or catharsis, or resort to neuroleptics.
3. *Encouraging suicidal patients to experience success and trust.* It is important, by both encouraging patients and analysing the reasons for their negative image of themselves and their living situation, to insert a positive chain of development into the therapeutic process. The feeling of "all or nothing" that the suicidal patient often has must be surmounted by the therapist and patient together, reaching an understanding of the patient's unrealistic goals and expectations and, by means of cautious explanations, clarifications and interpretations helping the patient to adjust to reality. Group therapy for suicidal patients may be successful. The patients then come into contact with others and, by perceiving their problems, can grow in terms of self-reflection and self-understanding.
4. *Stimulating the suicidal patient's imagination in a positive direction.* Suicidal thoughts arise in the absence of attractive plans for the future. If the suicidal patient starts to talk about future plans, this is an encouraging sign. It is then important to stimulate the patient's imagination by providing new ideas and asking about details. It is essential to remember that suicidal thoughts cannot be overcome by rational means. In terminating the treatment, it is important not to sever the ties abruptly. This means

that one has to follow up the patient, either personally or through support organisations.

Ringel was a proponent of psychotherapeutic treatment and thought that, in Vienna and in Austria as a whole, there were, in the 1970s, too few therapists in relation to the number of people in need of psychotherapy. The same situation applies today, in 1995, in Sweden.

In attempting to answer the question of what effect psychotherapeutic methods have on the treatment of depression and prevention of suicide, I would like to focus on the results of two recent studies. One of them deals with the treatment of depressed patients [3–6]. Since depression is closely associated with suicidality, it is helpful to scrutinize these results and learn from them. The second study deals with the treatment of borderline suicidal females [7–8].

II. National Institute of Mental Health (NIMH) Treatment of Depression Collaborative Research Program

The effectiveness of two brief psychotherapies, interpersonal psychotherapy and cognitive behaviour therapy, in the treatment of outpatients with major depressive disorder diagnosed by Research Diagnostic Criteria [9] was studied and compared with the use of a placebo and imipramine hydrochloride (antidepressant) treatment.

Interpersonal therapy and cognitive behaviour therapy are time-limited and manual-based psychotherapies.

A. Interpersonal Psychotherapy (IPT)

Interpersonal therapy was developed by Klerman et al. [10] in order to reduce some of the social problems that may provoke or prolong depression, such as loss of close relationships or conflicts in such relationships.

The therapist helps the patient recognize links between mood and current interpersonal experiences, focusing the therapy on one of four interpersonal problem areas: grief, role dispute, role transition and interpersonal deficits.

B. Cognitive Behaviour Therapy (CBT)

Cognitive behaviour therapy is a structured, time-limited psychotherapy developed by Beck and his colleagues [11]. The focus of cognitive behaviour therapy are the "automatic" negative thoughts that depressed patients report about themselves, their situations, and their future. Through rational discussions with their therapists and written and behavioural homework assignments, patients learn to modify irrationally negative thoughts; as they do so, their depression is alleviated.

III. Design of the NIMH Study of Depression

A. Acute 16-week Treatment

Two hundred and fifty patients were randomly assigned to four 16-week treatments: interpersonal psychotherapy, cognitive behaviour therapy, imipramine hydrochloride plus clinical management (as a standard reference treatment) or placebo plus clinical management.

Clinical management, in the NIMH study sometimes called medication clinic visits can be seen per se as a kind of supportive psychotherapy.

After evaluation of the acute treatment the authors proceeded with a maintenance project.

B. Maintenance Study

A randomized three-year maintenance trial was conducted in 128 patients with recurrent depression who had, in the acute 16-week study, responded to treatment with imipramine hydrochloride and interpersonal psychotherapy. A five-cell design was used.

The five treatments were: 1) a maintenance form of interpersonal psychotherapy (IPT-M) offered alone; 2) IPT-M with imipramine therapy continued at the acute treatment dosage; 3) IPT-M with placebo; 4) medication clinic visits with imipramine therapy; and 5) medication clinic visits with placebo.

IV. Results of the NIMH Study of Depression

A. Acute Treatment

Patients in all treatment groups, including placebo, during acute 16-week treatment showed significant reduction in depressive symptoms and improvement in functioning over the course of treatment.

Secondary analyses, in which patients were dichotomized according to their initial level of severity of depressive symptoms (HRSD severity criterion [12]) and impairment of functioning (GAS severity criterion [13]) showed that significant differences between treatments existed for the subgroup of patients who were more severely depressed and functionally impaired. There was some evidence of the effectiveness of interpersonal psychotherapy and strong evidence of the effectiveness of imipramine plus clinical management for functionally impaired patients. In the case of more severely depressed patients, IPT was as successful as treatment with imipramine plus clinical management. Interestingly, there were no significant differences between treatments, including placebo plus clinical management, for the less severely depressed and functionally impaired patients.

B. Results of Three-year Maintenance Study

Results of the three-year maintenance trial showed a highly significant prophylactic effect for interpersonal psychotherapy combined with imipramine hydrochloride treatment and for medication clinic with imipramine hydrochloride maintained at an average dose of 200 mg. A modest prophylactic effect for monthly interpersonal psychotherapy alone was also demonstrated. The authors concluded that active imipramine hydrochloride maintained at an average dose of 200 mg is an effective means of preventing recurrence of depression and that monthly interpersonal psychotherapy serves to lengthen the period between episodes in patients not receiving active medication.

V. Critical Evaluation of the NIMH Study of Depression

The highly ambitious and valuable NIMH study of depression taught us that several factors – such as randomization procedure, attrition/compliance, therapists' personality and skill, adherence to the manual, "dosages" of the therapies, placebo effects and number of patients in trials, as well as accounting in survival analyses for previous patterns of depression and suicidal behaviours – must be taken into consideration when evaluating the effects of psychotherapies and medication.

A. Randomization

1. Demographic Variables and Psychiatric Diagnoses

Critical evaluation of the study shows that there were some clearly observable effects of the randomization procedure. The randomization procedure seemed to be effective in creating groups that were equivalent with respect to demographic and baseline clinical characteristics but, unfortunately, not concerning definite endogenous diagnosis, which tended to be overrepresented in the two psychotherapy groups and in the placebo group.

The significant relationship to outcome, i.e. period before recurrence of depression, in a three-year maintenance study might be caused by a disproportionate number of subjects with a definite endogenous subtype having been assigned to the placebo or psychotherapy. Of the 76 subjects with a definite endogenous diagnosis, 51 (67.1%) were assigned to a no-medication treatment condition.

2. Personality Traits of the Patients

In the randomization procedure for different treatments, not only the demographic variables and the main psychiatric diagnoses, but also the patients' personality traits, should be taken into consideration. The capacity for benefiting from a particular treatment varies between different personality types.

Anaclitic patients – those who mainly have problems with disturbed relationships and who use evasive defences – improve more from psychotherapy with supportive elements than from insight psychotherapy.

Introjective patients, who focus more on their self-concept and use defences like projection and splitting, make better use of reflection-oriented methods.

The evaluation of psychotherapy outcomes in the above-mentioned project did not take into account the fact that the groups investigated may have been too heterogeneous, from the personality point of view, for the treatment method chosen.

3. Social Conditions and Family Situation

The social conditions of the patients and their family situation should also be taken into consideration in the randomization process. It is well known that the quality of the social and family network has an influence on the development of the suicidal process and on the treatment [14,15].

B. Attrition from and Compliance to the Treatment

Attrition during the three-year maintenance study was another problem. Twenty-two (17%) of the 128 patients assigned to the maintenance phase failed to complete the three-year protocol. There was a trend for greater attrition from imipramine cells and less attrition from the psychotherapy or no-pill cells. Attrition and compliance problems depend strongly on side-effects of the medication. In psychotherapy, side-effect problems are avoided, but on the other hand the therapist's personal qualities, involvement and skill and degree of adherence to the manual can significantly influence compliance with the treatment.

C. "Dosages" of the Treatment

The survival analysis in a three-year maintenance study showed that imipramine, at an average dose of 200 mg, provided good protection against recurrence of depression. For patients who did not receive medication, continued monthly IPT sessions significantly extended survival time, with a median survival time of more than one year.

In this context, it should be pointed out that while imipramine was given in doses of 200 mg/day, which is higher than that usually used for maintenance treatment, the "dose" of IPT once a month during maintenance was lower than in any previously published studies.

It may be asked whether the patients who were treated with psychotherapy alone would have experienced longer illness-free periods if the psychotherapy had been offered on a bi-weekly instead of monthly basis. Additional studies are needed to determine whether such more frequent IPT-M sessions might add to its preventive efficacy in patients not receiving maintenance pharmacotherapy.

Moreover, it should be underlined that continued treatment in the form of interpersonal psychotherapy together with imipramine hydrochloride showed clearly better results in terms of the recurrence of depression up to 60 weeks than imipramine treatment alone.

D. Placebo Effects

The effects of a placebo, as well as of the natural course of depression and of the suicidal process, should also be taken into consideration when evaluating the effects of different treatments.

An important feature of the NIMH trial was the inclusion of a condition combining a placebo and supportive interviews (clinical management).

The results of the NIMH study show that in acute 16-week treatment for less severely depressed patients, a placebo with clinical management had just as good an effect as treatment with antidepressants and with the two psychotherapeutic methods (IPT and CBT). In the patients who were more severely depressed, imipramine, IPT and CBT were significantly superior to a placebo.

It may be tempting to draw the conclusion that medical overtreatment presumably takes place in the moderately symptom-burdened cases of depression.

The placebo effect as such is known to account for one-third – and sometimes more – of a beneficial result. Both the natural course of an illness and the specific as well as unspecific doctor-patient relationship can account for the placebo effect. Factors accounting for the doctor-patient relationship include the patient's transference reactions and the doctor's attitudes towards them, in the form of the doctor's countertransference [2].

VI. NIMH's Five-year Outcome Study for Maintenance Therapies in Recurrent Depression

After conducting a randomized, three-year maintenance trial in 128 patients with recurrent depression, the authors asked the individuals who survived the three-year trial receiving active medication with or without psychotherapy to continue in a two-year additional randomized trial of active medication vs placebo.

The authors posed the question of whether maintaining antidepressant medication beyond three years at the dosage used to treat the acute episode would provide a prophylactic effect, compared with medication discontinuation after three years of maintenance treatment.

VII. Critical Evaluation

A. Small Number of Patients and Disregard of Accounting for Pre-trial Course of Depression and Suicidal Process in Performing Survival Analyses

Survival analysis performed on 11 patients treated with imipramine hydrochloride and 9 patients treated with a placebo showed a significant prophylactic effect for imipramine hydrochloride treatment maintained at an average dose of 200 mg. In spite of the extremely low number of patients included in the study the authors concluded that patients with previous episodes of depression less than 22 years apart should continue prophylaxis and be treated with antidepressants for at least 5 years.

Both the prophylaxis provided by medication and that provided by psychotherapy need to be viewed in the context of the anamneses of the patients studied. The intensity and time variation of previous depressions and suicide attempts should be included in the analyses in order to minimize, in statistical calculations of survival time, the influence of the natural course of illness.

VIII. Dialectic Behavioural Therapy (DBT)

The only specific study of therapeutic treatment of suicidal patients is a well-documented trial by Linehan [7, 8]. In this study a Dialectic Behavioural Therapy (DBT) was used. DBT increases behavioural skills by mindfulness of core skills, affect regulation, distress tolerance, self-management and appropriate handling of interpersonal relations. In DBT, active individual and group therapy are combined. The therapists work on conscious material and do not deliberately work on suicidal patient's unconscious fantasies and conflicts.

Linehan's study shows good results for female borderline patients with lasting suicidal tendencies who were treated for a year with DBT compared with a control group who received "treatment as usual". The patients who had received DBT had fewer suicide attempts and fewer days' inpatient care during the year of treatment. However, no follow-up study has been reported, and the number of patients in the study was also very small.

This study is subject to the same criticism concerning randomization procedure, attrition/compliance and number of patients as the NIMH study of depression analysed above.

A critical examination of the results from both the NIMH study and the Linehan study makes it clear that, in the random assignment of patients to different treatment groups, it is extremely important to take into account not only demographic and psychiatric diagnoses, but also other diagnostic variables (psychological, social and family situation).

In evaluating the preventive effects of different therapies it is also important to consider the intensity of the suicidal process, i.e., the number and timing of previous suicide attempts, before the patient joins the project. Statistical analyses should be carried out so that in survival analyses, i.e., assessments of the probable number of future repeated suicide attempts or suicides, the previous suicide-attempt pattern, i.e., the pattern before patients enter the study, is taken into account. Survival analyses of suicidal patients performed at our centre in which gender, age, marital status, personality type and psychiatric diagnoses according to DSM-III-R, Axis I-V, and also previous suicide-attempt patterns, were taken into account, show that the recurrence of suicide attempts varies appreciably between different groups of patients. Linehan's results should therefore be re-calculated both for patients treated with DBT and for those who received "treatment as usual".

At the Karolinska Institute in Stockholm, a psychotherapy project is planned for borderline women with suicidal tendencies, who are to be treated with (a) Dialectic Behavioural Therapy in co-operation with Marsha Linehan and (b) Psychodynamic Therapy for Borderline Patients in co-operation with Otto Kernberg. The psychotherapy project will be headed by M. Åsberg and the evaluation of suicide-preventive effects of this project directed by the author.

In the evaluation of suicide-preventive effects of the psychotherapies used in the Stockholm project, besides epidemiological analyses taking into consideration aspects discussed in this article, the importance of the therapists' attitudes towards the suicidal patients will also be assessed. All tape-recorded and videotaped sessions will be evaluated with a quantitative instrument devised by psychoanalyst Paulina Kernberg at the New York Hospital – Cornell Medical Center, Westchester Division in New York. For comparison in the psychotherapy project, inclusion of a group of suicidal women treated with antidepressants or lithium is also planned. If such patients are included in the psychotherapy project, doctors participating in this part of the study will also be assessed with P. Kernberg's instrument. With this quantitative method, the author would like to further develop previous results obtained by qualitative methodology on transference and countertransference aspects in the treatment of suicidal patients [16]. The hypothesis that the therapist's qualities are probably just as important in the treatment of suicidal patients as the framework and content of the therapy itself will be tested.

IX. Conclusions

Drug treatment has several shortcomings. Some patients fail to respond to an adequate dosage and side-effects may be troublesome. These problems can be reduced by using the newer antidepressants or psychotherapeutic methods. From the results of the NIMH study it appears that for the patient who cannot, or chooses not to, continue receiving pharmacotherapy, even monthly sessions of Interpersonal Psychotherapy (IPT) may be helpful in extending the period between depressive episodes. This finding is of particular importance in the case of the woman wishing to become pregnant or of the patient who has a medical condition incompatible with tricyclic pharmacotherapy. The reported positive effects of Dialectic Behavioural Therapy (DBT) in the treatment of suicidal patients and the reported cumulative effect of antidepressants and interpersonal therapy in the treatment of depression make it reasonable to try psychological treatment for patients whose relapse in depression and/or suicide attempts seems to be related to cognitive factors or interpersonal problems.

In conclusion, it is my opinion that Erwin Ringel was right in believing that psychotherapy is beneficial for many suicidal patients. Psychotherapy has the advantage of empowering patients to prove to themselves that they can control their mood and their environment. However, medication may provide faster relief and offer more reliable prophylaxis against recurrence in patients with severe depression. It should also be underlined that psychotherapy undoubtedly augments the benefits of antidepressant medication for medication responders. Therefore the author strongly recommends combining antidepressants medication and psychotherapy in the treatment of severely depressed patients.

References

1. Ringel E. Selbstmordverhütung. Verlag Hans Huber, Bern, 1969.
2. Wolk-Wasserman D. Some problems connected with the treatment of suicide attempt patients: Transference and countertransference aspects. Crisis, 1987; 8/1: 69–82.
3. Elkin I, Shea MT, et al. National Institute's of Mental Health Collaborative Research Program on Treatment of Depression. General Effectiveness of Treatments. Arch Gen Psychiatry, 1989; 46: 971–982.
4. Frank E, Kupfer D, et al. Three-year outcomes for maintenance therapies in recurrent depression. Arch Gen Psychiatry, 1990; 47: 1093–1099.
5. Frank E. Interpersonal psychotherapy as a maintenance treatment for patients with recurrent depression. Psychotherapy, 1991; 2: 259–266.
6. Kupfer D, Frank E, et al. Five-year outcome for maintenance therapies in recurrent depression. Arch Gen Psychiatry, 1992; 49: 769–773.
7. Linehan MM, Armstrong HE, Suarez A, Allmon D, Heard HL. Cognitive-behavioral treatment of chronically parasuicidal borderline patients. Arch Gen Psychiatry, 1991; 48: 1060–1064.
8. Linehan MM, Heard HL, Armstrong HE. Naturalistic follow-up of a behavioral treatment for chronically parasuicidal borderline patients. Arch Gen Psychiatry, 1993; 50: 971–974.
9. Spitzer RL, Endicott J, Robins E. Research Diagnostic Criteria: rationale and reliability. Arch Gen Psychiatry, 1978; 35: 773–782.
10. Klerman GL, Weissman MM, Rounsaville BJ, Chevron ES. Interpersonal Psychotherapy of Depression. New York, NY: Basic Books Inc Publishers, 1984.
11. Beck AT, Rush AJ, Shaw BF, Emery G. Cognitive Therapy of Depression. New York, NY: Guilford Press, 1979.
12. Hamilton MA. Development of a rating scale for primary depressive illness. Br J Soc Clin Psychol, 1967; 6: 278–296.

13. Endicott J, Spitzer RL, Fleiss JL, Cohen J. The Global Assessment Scale: a procedure for measuring overall severity of psychiatric disturbance. Arch Gen Psychiatry, 1976; 33: 766–771.
14. Wolk-Wasserman D. Suicidal communication of persons attempting suicide and responses of significant others. Acta Psychiatr Scand, 1986; 73: 481–499.
15. Wasserman D. Passive euthanasia in response to attempted suicide: one form of aggressiveness by relatives. Acta Psychiatr Scand, 1989; 79: 460–467.
16. Wolk-Wasserman D. Attempted suicide – the patient's family, social network and therapy. Doctoral dissertation. Stockholm: Graphic Systems, 1986: 1–220.

20. Applications of Solution-Focused Brief Therapy in Suicide Prevention

HEATHER FISKE

Solution-focused brief therapy (SFBT) has been described as part of a "megatrend" in psychotherapy in which the focus of treatment has shifted "away from explanations, problems, and pathology, and toward solutions, competence, and capabilities" (O'Hanlon & Weiner-Davis, 1989, p.6). SFBT has much in common with other, earlier models as well – for example, with the optimism and fostering of self-efficacy in Ringel's Adlerian approach to suicide prevention (Diekstra, 1995). The approach has been developed by Insoo Kim Berg, Steve de Shazer and their colleagues at the Brief Family Therapy Center in Milwaukee (de Shazer 1988, 1994; de Shazer, Berg, Lipchik, Nunnally, Molnar, Gingerich, & Weiner-Davis, 1986; Kral & Kowalski, 1989; Nunnally, de Shazer, Lipchik, & Berg, 1987; Walters & Peller, 1992; Weiner-Davis, de Shazer, & Gingerich, 1987). Applications in a wide variety of settings and populations have had positive results (e.g. Berg, 1994a; Berg & Miller, 1994; Booker & Blymer, 1994; Dolan, 1992; Ingersoll-Dayton & Rader, 1993; Kral & Schaffer, 1989; Mc Farland, 1995; Peller & Walters, 1989; Plaxton, 1995; Weiner-Davis, 1987).

SFBT is a respectful and empowering approach which in my view has much to offer in clinical suicide prevention. In particular, solution-focused work may allow helpers to begin the "search for solutions" immediately, even during crisis intervention. In addition, solution-focused techniques may provide tools for decreasing perceptual constriction, working with ambivalence, facilitating early communication of intent, and developing workable alternatives to suicide which are consistent with individual needs.

In this paper I will briefly outline the central philosophy and assumptions of SFBT, and describe some of the associated therapeutic techniques which may be helpful in clinical suicide prevention work. As a general framework for discussing SFBT applications, I will use Shneidman's "Ten commonalities of suicide" (Shneidman, 1987, 1989). The "ten commonalities" are presented here, not as a definitive or comprehensive description of the suicidal person, but as a vehicle for illustrating some typical concerns and issues in suicide prevention. My intention in choosing this format is to give interested clinicians

D. De Leo, A. Schmidtke and R.F.W. Diekstra (eds.), Suicide Prevention, 185–197.

a sense of how this model can be applied in addressing such concerns and issues.

I. SFBT Philosophy and Assumptions

A. The Central Philosophy

Berg and de Shazer have expressed the Central Philosophy of Solution-Focused Brief Therapy in three succinct "rules of thumb":

Rule l: If it isn't broken, don't fix it.

Rule 2: If it's working, do more of it.

Rule 3: If it's not working – *do something else.*

These guidelines can provide a pragmatic standard against which helping professionals can evaluate their clinical work on a continuing basis:

- Am I attending to the problem presented, or getting off course, following my own agenda? (Rule 1)
- How can I maximize what I am doing that is effective? (Rule 2)
- What can I do differently instead of what is not working? (Rule 3)

In crisis intervention situations, there may be multiple problems, multiple options, and limited time in which to evaluate the information. Reference to the Central Philosophy can help interveners to focus, and stay focused, on the most effective course of action.

B. Assumptions

Talking solutions.

A fundamental assumption of solution-focused work is that "Focusing on the positive, the solution, and the future facilitates change in the desired direction (Walters & Peller, 1992, p.10). As a result, solution-oriented talk is seen as a more valuable focus than problem-oriented talk. This shift in focus is a profound change for most traditionally-trained mental health professionals – but it is only a shift, not a complete departure. Problem definition is still an important part of a solution-focused therapeutic process; inviting clients to tell us their stories and listening reflectively remain fundamental. Careful listening and clinical judgement help to determine the most appropriate timing for introducing "solution talk".

One of the key aspects of "solution talk" is the emphasis on *exceptions* to the problem, on those times when something other than the problem is occurring. Such exceptions invariably occur, even when the problems seem particularly pervasive: as de Shazer says, "nothing always happens". In solution-focused therapy, it is assumed that exceptions to the problem can be used as a foundation to build solutions.

Utilization.

The principle of *utilization*, derived from the work of Milton Erickson, suggests that we can help people most effectively by utilizing their own competencies, strengths, resources and successes. Implicit within this principle is the basic assumption that people *have* competencies, strengths, resources and successes; and further, that what they have includes whatever they need to solve their problems.

Client as expert.

In solution-focused work, the goals and solutions, as well as the utilizable resources, are provided by the client. Rather than fitting the client's problems and characteristics, needs and goals, into a series of pre-existing categories defined by the therapist's expertise, it is the therapist's task to help clients define *their* goals as clearly and concretely as possible. Berg and de Shazer refer to the therapeutic stance which fosters this kind of helpfulness as a posture of "not knowing".

Nature of change.

There are several solution-focused assumptions regarding the nature of change. One of these is that change is constantly occurring; nothing (and no problem) is stable, infinite, or inevitable. A second assumption is that change is *generative*; i.e., small changes lead to larger changes. One corollary of this second assumption about change is that beginning with small, workable, "everyday" goals for change is a powerful first step toward more profound and pervasive change. A third assumption regarding change is that clients show us through their behaviour how *they* believe change takes place. According to this assumption, clients are always "cooperating" with us in the effort to change their lives for the better. If clients seem "resistant" to us, it is because we have not yet understood their model of change, nor learned to cooperate with it.

To summarize: SFBT is a goal- or solution-focused endeavour with the client as expert. The therapist's role is to facilitate recognition and implementation of goals and solutions:

> The competent clinician listens carefully to her clients with respect and curiousity, recognizing that there are multiple views of every event. She will understand that therapy is not to dissolve problems....Rather it is a context the therapist offers the client for evolving new meanings, new ways of looking at the vexing problems that brought him to the therapist. The therapist, through raising questions, opens up the possibility for new ways of looking at problems. This is the beginning of solution building. The therapist takes a posture of "not knowing", a non-hierarchical and facilitative role, thus making it possible for the client to create a solution that is congruent with his way of conducting life. (Berg, 1994b, p.13).

II. SFBT Techniques

A number of useful techniques have been developed within the framework of solution-focused assumptions and are frequently used by solution-focused therapists. On that basis, they can be described as "solution-focused techniques"; some of them will be described here. However, the most important criterion for decisions about technique from a solution-focused perspective is the Central Philosophy. So, for example, I continue to incorporate "no-suicide decisions" (Drye, Goulding, & Goulding, 1973) in my "solution-focused" clinical work because I often find them to be useful (see Rule 2). The question for helping professionals is one of "goodness of fit": What in this approach can be integrated with how I work as a therapist, in my working environment, with my clients, to make my work more helpful?

A. Useful Questions

Many of the solution-focused techniques take the form of questions, which is consistent with the nonnormative, individualized nature of the approach. In the opening stages, typical questions invite the client to describe their problems and goals:

- What brings you in today?
- How can this meeting be helpful to you?
- What can happen here today in order for you to know that it was a good idea to come?

Another kind of question typically asked early in the first session is *questions about pre-session change*, for example:

- What has been different since you made the decision/ booked the appointment to come here?

In asking such questions, therapists recognize that clients are already "doing something different" by picking up the telephone or walking into a crisis unit to ask for help. In at least two-thirds of cases, clients, *if asked*, will describe positive changes – changes which can be used to construct larger solutions (Berg & de Shazer, 1994)

One of the most powerful change-generating techniques developed by Berg and de Shazer is the *miracle question*, which establishes a hypothetical "solution picture" for each client. This question takes the following general form:

- Suppose that, after our discussion today, you leave here and go home. Tonight, while you are sleeping, a miracle happens. As a result of this miracle, the problem that has brought you here today is completely solved. Because you are sleeping, you don't know that a miracle has happened and the problem is solved. How will you find out? What will you notice tomorrow that will tell you that there has been a miracle?

As the therapist continues to inquire about details of the "miracle picture", the projected solution becomes increasingly "alive" for both therapist and client.

Clients' responses to the miracle question (and other questions as well) can be expanded and developed through the use of *relationship questions*, which enhance the detailing of the solution picture and place it in the context of the person' s social environment:

- What will your wife notice that will tell her that a miracle has happened? What will she be doing differently? What will you be able to do that you are not doing now, as a result of her changes ? (etc.)
- What about the people at work? at school? What will they notice?
- Following the miracle question, *questions about exceptions* to the problem can readily be asked in the following way:
- When is even a small part of your "miracle" happening already?
- What difference does that make for you?
- How are you doing that?

Quantifying questions, which use scales or percentages to define where people see themselves in the process of change, are helpful in a variety of ways (Berg & de Shazer, 1993, 1994).

First, these scales can be used to construct criteria for improvement, and to measure it as it occurs:

- On a scale of 1 to 10, where are you now (with regard to the specific problem) ?
- Where do you want to be?
- What would you need to do in order to move up one point on the scale?

Similarly, scaling can be used to measure a variety of salient aspects of the therapeutic process – for example, motivation:

- On a scale of 1 to 10, if 10 is "I would do anything" and 1 is "I wouldn't lift a finger", how willing are you to do whatever is needed to solve this problem?
- What small step can you take that would increase your willingness?
- How far will doing that move you on the scale?

Percentage questions are helpful in similar ways – for example, in defining exceptions to the problem:

- What percentage of the time does this problem occur?
- ...not occur?
- What percentage would be normal (to be expected)?

Advantages of quantifying questions are that they can be used even with relatively non-verbal clients, and can be presented in a variety of formats: as graphs, pictures, concrete representations, or in physical movement.

At the beginning of second and later sessions, positive change can be elicited by asking:

- What has been better since we last met?
- or,

– What has been different?

At any point, the number of positive changes, or the details of goals and solutions, can be amplified and expanded by asking:

– And what else?

With clients who are extremely negative, hopeless, and pessimistic, *coping questions* may be very helpful. These questions are respectful of the client's position but focus on the person's capacity for health and survival:

– Why aren't things worse? With all you have been through, how have you managed to keep things from getting any worse?
– How have you coped up to now? What have you been doing to get by?

B. Use of language

One of the featuress of solution-focused therapy as it is developing at the Brief Family Therapy Center, is a recognition that therapy sessions can be seen as "nothing...but an exchange of words" (Freud, 1915, quoted in de Shazer, 1994). Thus, careful attention to the use(s) and impact of language in therapy is an integral part of this approach.

Characteristic of the model is the use of *presuppositional language*, which can be used to convey, for example, assumptions and expectations of agency or of positive outcome:

– "How did you do that?" rather than "How did that happen?"
– "What will be different when you are feeling better?" rather than "What would be different if you were feeling better?"

C. Feedback

There are two primary kinds of feedback which are typical of solution-focused work: *task assignments* and *compliments*. Tasks may vary from observation assignments (e.g. "notice when small parts of the miracle picture are already happening and what is different about those times") to more active goal-oriented "experiments" (e.g. "smile at your boss every other day next week and observe what happens"). It is important that tasks are congruent with the level of motivation and therapeutic relationship (Berg, 1989, Berg & Miller, 1993), and that task assignments be accompanied by a "rationale" which describes the purpose and function of the task in language that is meaningful to the client.

From a solution-focused perspective, compliments are an important and powerful clinical intervention. Compliments are most effective in orienting people to their own strengths, resources, competencies, and successes under the following conditions: when they refer to the client's behaviour; are sincere; affirm how difficult the client's problems are; and reinforce positive coping.

"Indirect" compliments are also very helpful:

– Wow! Really? How did you manage that?

III. Commonalities of Suicide: SFBT Applications

In this section, Shneidman's ten commonalities are listed. Each is accompanied by a brief discussion of how solution-focused brief therapy might help to address the needs and challenges it represents.

1. The common purpose of suicide is to seek a solution
Shneidman views every suicidal person as seeking solutions. His first commonality fits very well with the solution-focused premises that people are constantly changing, and that they are always, in some way, cooperating with our attempts to help them. It is the clinician's task to recognize individual processes of change and cooperation, and to work with those processes. Understanding that the suicidal person's behaviour is oriented toward solving life problems is a first step in allowing the clinician to adopt a collaborative, mutual stance with the individual. Understanding *how* the individual views suicide as a personal solution is a first step toward understanding what *else* could serve as a solution for this person.

2. The common goal of suicide is cessation of consciousness
In a solution-focused approach, the clinician recognizes goal-directedness even in "negative" behaviour. Without reinforcing the particular goal, it is then possible to work with the person "*where the person is*", and to begin to understand what their goals mean to them. Questions such as the following may be helpful:

- How would that (i.e. achieving your goal) be helpful to you?
- What else would be different/better?

Once the desired consequences of suicidal behaviour are understood, it is possible to consider alternatives to suicide which may achieve similar consequences in the person's life:

- When were you doing even a little bit better?
  How did you figure that out?
- How can you do more of that?
- When you are doing more of that, how will things be different? What is the first thing you will notice?

3. The common stimulus in suicide is intolerable psychological pain
First, we must accept the reality of the person's pain as the person perceives it. Second, when suicide is viewed as an escape from pain, anything which helps to decrease the pain even slightly can be utilized in reducing suicide risk. In this context, "anything" may – and often does – include thoughts and activities which we view as undesirable or unhealthy.

To illustrate: some years ago, I was treating a young woman who had been diagnosed as bulimic and who was going through a period of severe depression. She told me one day that she had very much wanted to kill herself the day before but had not because she "couldn't bear to die fat" (she was in fact slightly underweight). Her negative body image and her desperate hope that she might yet achieve physical "perfection" were keeping her alive.

Yvonne Dolan (1992) has described a solution-focused sequence in which "negative" pain-reducing alternatives such as non-lethal self-harm, purging, substance abuse, etc., are examined in imagination. The individual is asked to notice how, and how much, each of these imagined behaviours reduces the emotional pain. As a scenario develops in which the individual has been able to act effectively in reducing the pain, the clinician continues to ask about further alternatives for reducing the pain ("what else?"). Eventually, they may have composed a surprisingly long list, and the list is likely to include some non-destructive coping mechanisms.

An awareness of the impact that small changes can make and the facilitation of a "just noticeable difference" are also useful in working with people in pain. Quantitative questions (e.g. scaling) are often helpful in articulating the parameters both of the person's pain and of some measure of control over it:

- On a scale of 1 to 10, if 1 is no pain and 10 is the worst pain you ever felt, where are you today?
- What is the best you have been in the last day?/ week?/ month?
- What was different? What were you doing to decrease the pain – even a little bit?
- If your pain is at 9 on the scale, what could you do to decrease it to 8.5? How will you notice this difference?

4. The common stressor in suicide is frustrated psychological needs

Central to a solution-focused approach is an understanding of the meaning of the unmet needs *for this person*. In SFBT, the unmet needs are translated from "what I'm not getting" to "what I need and what my life will look like when I get those things". The impact of this "translation" is to shift the focus from the problem state to the goal picture. In most cases, people's goals, when described in detail and in positive terms, are ordinary, human, and workable. The miracle question is particularly helpful in this regard. Surprisingly, the most common first response to the miracle question does not involve radical or unusual events; according to research at the Brief Family Therapy Center, what people most often say is that one of their family members would be smiling at them (Berg & de Shazer, 1994).

Another common type of response to the miracle question is that something negative will be absent–e.g., "My parents won't be yelling at me", "I won't feel so hopeless". This kind of response can be restated in positive terms that allow for positive goal formulation and exception-finding by asking:

- what will be happening *instead* of [the negative event]?

5. The common emotion in suicide is helplessness-hopelessness

The central contribution of solution-focused therapy in dealing with feelings of helplessness-hopelessness is the orientation to *exceptions* – to what the suicidal person *can* and *does* do, despite what may seem to that person, and sometimes to the clinician as well, to be overwhelming problems. One example of this orientation in the suicide prevention literature is Rotheram's (1987) model for evaluating "imminent danger" for suicide among runaway youth (a group at relatively high risk for suicide). In triage programs with this population, Rotheram found that the assessment format which was most effective in differentiating those in imminent danger of suicide included concrete measures of the individuals' ability to behave in a non-suicidal manner, as well as a more traditional statistically-based evaluation of suicide risk. Berman and Jobes (1991) also recommend including an evaluation of coping skills and resources as part of suicide risk assessment.

A similar combination of questions about dimensions of suicide risk and questions about more hopeful and efficacious behaviours may be part of a solution-focused suicide interview (Brief Family Therapy Center, 1990):

- Previous attempts?
- What have you done to survive long enough to get here?
- Frequency of thoughts of dying or suicide?
- When these thoughts come, how long do they last?
- What are you doing to make these thoughts go away?

The use of presuppositional language in these solution-focused questions conveys implicit assumptions of action, efficacy, and hope, such as: you have done things to assist your own survival; thoughts of suicide do not last indefinitely; you have the ability to modify these thoughts.

6. The common cognitive state in suicide is ambivalence

Ambivalence is a major challenge in clinical work with suicidal individuals: How do we recognize and support the desire to live without trivializing the person's pain and distress? The Reasons for Living Inventory (Linehan, Goodstein, Nielsen, & Chiles 1983) is one attempt to address this question.

Solution-focused techniques offer tools for helping to articulate both sides of the ambivalence. Once the more life-affirming aspects have been recognized, the exceptions which support them can be evoked. Again, quantifying questions may be especially helpful:

- What percentage of the time do you want to kill yourself?
- If you want to kill yourself 90%' of the time, what about that other 10%? What are you doing differently at those times?
- What could happen to increase that percentage by even 5%?
- On a scale of 1 to 10, how likely are you to do some of the things that will increase that percentage?
- What could make you even a little bit more likely to do those things?

7. The common perceptual state in suicide is constriction
Toward the end of the movie "Schindler's List", there is a scene which illustrates an effective interruption of perceptual constriction. World War II is ending, and Schindler is taking his leave of the workers whom he has saved from the concentration camps by pretending that they were necessary in his munitions factory. As he does so, he is overwhelmed by an awareness of how many more he could have saved, had he only done more, given more. He becomes frenzied in his guilt and self-accusation. At that point, the foreman touches his arm and insists that he looks at what he *has* done: "Look at us! We're alive because of you."

Insoo Kim Berg has described solution-focused questions as serving a function similar to the foreman's intervention with Schindler. She sees these questions acting as a "tap on the shoulder" to clients (Berg & de Shazer, 1994), redirecting their attention from a preoccupation with failure and disaster to a consideration of their own accomplishments, strengths, and resources, and to possibilities for a positive future.

8. The common interpersonal act in suicide is communication of intention
The communication of suicidal intent is common but by no means universal, even among those with access to helping professionals (e.g. Runeson, 1995, Groholt & Ekeberg, 1995). Further, the communication is not always conveyed in a "language" understood by the intended recipient at the time. I would contend that the communication of suicidal intent may be facilitated by solution-focused interactions. This facilitation occurs because of several aspects of the therapy context which are likely to promote such communication. First, clients quickly realize that the therapist will consider them as whole persons, focusing on their more healthy and positive attributes as well as on psychiatric symptoms and plans for suicide. Second, they see that the therapeutic emphasis is on finding concrete solutions and relief *in their terms*. Lastly, since clients are seen as the experts, their statements will be accepted and believed, rather than denied or minimized.

I would also contend that helping professionals working within a solution-focused framework may find it easier to ask about suicidal intent, and that they may be less likely to panic when it is articulated. While this contention may seem somewhat paradoxical – asking about suicide is certainly problem-focused-helping clients to tell their stories is an essential aspect of SFBT, and intent to commit suicide is likely to be an important element in anyone's story. Also, both the collaborative nature of the helping relationship and the clear sharing of responsibility may alleviate some of the burden felt by many clinicians in dealing with their clients' suicidal behaviours.

9. The common action in suicide is egression
When the desire to exit a painful situation is being enacted in suicidal behaviour, solution-focused methods provide a route to finding alternative – and

usually more palatable – goals. For example, one of my colleagues recently interviewed a family in which the adolescent daughter had been referred following a series of suicide attempts. When asked about her goals for treatment, the girl said that she wanted her parents' permission to kill herself. The therapist asked the girl (without condoning her request) how this would be better for her; the girl's reply was that granting permission for her suicide would mean that her parents had listened to her and had taken her distress seriously. They went on to discuss signs (other than permission to die) which would indicate that her parents were listening; when these signs were already present, even to a limited extent; and what signs her parents could look for from her to tell them that listening was working. Her complaint had shifted from not getting permission for suicide to not being listened to; the focus and goals of therapy had shifted from her desire to kill herself to how the family could communicate more effectively.

10. The common consistency in suicide is with lifelong coping patterns
From a solution-focused point of view, people are demonstrating coping skills even in crisis states. The focus on observing and utilizing strengths and resources that is such an integral part of the model can be particularly effective at such times. The solution-focused clinician often asks questions in the general format:

– How have you managed to _____ even in the midst of _____?.

Observations of coping strategies in the crisis situation can become the basis for recognizing the resources that will contribute to *non-suicidal* coping – for example:

> So... you've gone about this [plan for suicide) in a very determined way. Is that something you knew about yourself...that you are determined? In what other ways is this evident in your life? Is this something that [significant other] knows about you? In what ways does this person see your determination?

The solution-focused clinician's assumption is that people can lcarn from and rely on their own accomplishments, even in the face of terrible pain and fear and apathy.

IV. Conclusions

It is important to note that while the fundamental ideas of solution-focused therapy are simple ones, applying them in a simplistic or formulaic way is directly contrary to the tenets of the model. Solution-focused techniques "are not a bag of tricks; the questions articulated by solution-focused therapists are expressions of an attitude, a posture, and a philosophy" (Berg, 1994b). However, applied with care and respect, these methods can do a great deal to relieve pain and perturbation in suicidal people, and to help them reclaim worthwhile lives.

To illustrate: I was recently involved in a case in which an innovative suicide prevention plan, perfectly tailored for the individual, was developed – not by me, but by the client. This young woman was becoming increasingly depressed and preoccupied with suicide. She was particularly frightened by her experience because in a previous episode of depression she had continued to feel this way for almost a year; she did not feel that she could endure that again, and saw suicide as preferable. I asked her what she had learned from that earlier time in her life. She answered that she didn't know, all she did then was lie on the couch, sleep, and read horror novels. I asked how this had helped her to cope, and she replied that while she was reading, she had been safe because "I couldn't bear to kill myself without knowing how the book turned out". Finally, I asked her how this hard-won knowledge could help her now. After a long silence, she replied that what she was going to do was to *write* a horror novel; this would take at least a year, she wouldn't be able to kill herself until it was finished, and by then "something will have changed".

With all due respect for my own clinical acumen, I could never have created such a fitting solution – nor convinced her to implement it if I had. As I work more and more in a solution-focused way, I have more and more "surprises" of this kind. More and more, I am confronted by concrete evidence of the remarkable resources for hope and healing that exist in all of us.

References

Berg IK. Of visitors, complainants, and customers: Is there really any such thing as resistance. The Family Therapy Networker, 1989; 13(1): 21.

Berg IK. Family-based services. New York: Norton, 1994a.

Berg IK. A wolf in disguise is not a grandmother. Journal of Systemic Therapies, 1994b; 13(1): 13–14.

Berg IK, Miller, S. Working with the problem drinker: A solution-focused approach. New York: Norton, 1993.

Berg IK, de Shazer S. Making numbers talk: Language in therapy. In S Friedman (editor), The new language of change. New York: WW Norton, 1993.

Berg IK, de Shazer S. A tap on the shoulder: Six useful questions in building solutions. [Audiotape). Milwaukee, Wis. Brief Family Therapy Center, 1994.

Berman A, Jobes DA. Adolescent suicide: Assessment and intervention. Washington, D.C.. American Psychological Association, 1991.

de Shazer S. Clues: Investigating solutions in brief therapy. New York: Norton, 1988.

de Shazer S. Words were originally magic. New York: Norton, 1994.

de Shazer S, Berg IK, Lipchik E, Nunnally E, Molnar A, Gingerich W, Weiner-Davis M. Brief therapy: Focused solution development. Family Process, 1986; 25: 207–222.

Dolan Y. Resolving sexual abuse: Solution-focused therapy and Ericksonian hypnosis for adult survivors. New York: Norton, 1992.

Drye RC, Goulding RL, Goulding ME. No-suicide decisions: Patient monitoring of suicide risk. American Journal of Psychiatry, 1973; 130: 171–174.

Groholt B, Ekeberg O. Suicide among persons under 20 years. Presented at the Eighteenth Congress of the International Association for Suicide Prevention, Venice, Italy, 1995.
Ingersoll-Dayton B, Rader J. Searching for solutions: Mental health consultation in nursing homes. Clinical Gerontologist, 1993; 13(1): 33–50.
Kral R, Kowalski K. After the miracle: The second stage in Solution Focused Brief Therapy. Journal of Strategic and Systemic Therapies, 1989; 8(2): 73–76.
Kral R, Schaffer J. Treating the adoptive family. In CS Chilman, EW Nunnally, FM Cox (edsitors), Variant family forms: Families in trouble. Newbury Park: Sage, 1989.
Linehan MM, Goodstein JL, Nielsen SL, Chiles JA. Reasons for staying alive when you are thinking of killing yourself: The reasons for living inventory. Journal of Consulting and Clinical Psychology, 1983; 51, 276–286.
McFarland B. Brief therapy and eating disorders: A practical guide to solution-focused work with clients. San Francisco: Jossey-Bass, 1995.
Nunnally E, de Shazer S, Lipchik E, Berg I. A study of change: Therapeutic theory in process. In DE Efron (editor), Journeys: Expansion of the strategic-systemic therapies. New York: Brunner/Mazel, 1987.
O'Hanlon W, Wilk J. Shifting contexts: The generation of effective psychotherapy. New York: Guilford, 1987.
Peller J, Walters J. When doesn't the problem happen? In M Yapko (editor), Brief therapy approaches to treating anxiety and depression. New York: Brunner-Mazel, 1989.
Plaxton M. Introducing solution-focused therapy to inpatient psychiatric staff: A health-oriented approach. Unpublished manuscript, University of Toronto, Toronto, 1995.
Rotheram MJ. Evaluation of imminent danger for suicide among youth. American Journal of Orthopsychiatry, 1987; 57: 102–110.
Runeson BS. The suicidal process in young suicides. Presented at the Eighteenth Congress of the International Association for Suicide Prevention, Venice, Italy, 1995.
Shneidman ES. Definition of suicide. New York: John Wiley, 1985.
Shneidman ES. Overview: A multidimensional approach to suicide. In Douglas Jacobs & Herbert Brown (editors), Suicide-understanding and responding: Harvard Medical School perspectives. Madison, Conn.. International Universities Press, 1989.
Walters J, Peller J. Becoming solution-focused in brief therapy. New York: Brunner-Mazel, 1992.
Weiner-Davis M. Case studies: Confessions of an unabashed marriage-saver. The Family Therapy Networker, 1987; 11(1): 53–58.
Weiner-Davis M, de Shazer S, Gingerich W. Building on pretreatment change to construct the therapeutic solution: An exploratory study. Journal of Marital and Family Therapy, 1987; 13: 359–364.

21. Pharmacological Treatment of Suicidal Behavior

FABRIZIO SCHIFANO and DIEGO DE LEO

I. Introduction

One of the most important problems the clinician has to cope with is the prevention of suicide. In fact, about half of the people who commit suicide have seen a doctor in the few weeks prior to their death and often use prescribed medications to end their lives (Barraclough et al. 1974). It has been suggested that physicians see at least six seriously suicidal patients each year, but only one is detected, since the psychiatric diagnosis has been overlooked or the illness undertreated (Murphy, 1972; Oyehagen, 1991). Given the frequency with which psychotropic drugs are prescribed (mostly by general practitioners and not by specialists) in this therapeutic area, it seems worthwhile to review the pharmacological strategies aimed at preventing suicidal behavior.

II. Biochemical Factors

Many researchers have discussed the implications of serotonin (5-HT) metabolism in depression and suicidal behavior (Asberg et al. 1976; De Leo & Marazziti, 1988; Traskman et al. 1981) and there is now persuasive evidence that 5-HT system function is diminished in patients who exhibit suicidal behavior (Cohen et al. 1988; Rao et al. 1994). It is worth noting that low levels of cerebrospinal fluid 5-hydroxyindolacetic acid (5-HIAA, the metabolite of 5-HT) in depressed patients are more closely correlated with disturbed aggression regulation (i.e. violent suicidal attempts and outwardly directed aggression) than with depressive mood per se (Van Praag, 1986). The presence of contributory factors (hopelessness, impulsiveness, inadequate social support, a diminished central 5-HT turnover, family history of suicidal behavior) obviously increases the risk of suicide and the proposed model helps the physician to understand why not all patients suffering from affective disorders commit suicide: the sub-groups showing an overlap of several risk domains are at particular risk (Blumenthal, 1988).

D. De Leo, A. Schmidtke and R.F.W. Diekstra (eds.). Suicide Prevention, 199–209.

III. Pharmacological Intervention

A growing interest has recently been directed toward assessment of the preventive effect of drugs on suicidal behavior (Schifano, 1994). Bearing in mind the "low cost" of such treatments, the following is an up-date on their effectiveness with some practical recommendations on the use of antidepressants (ADs), neuroleptics (NLs), benzodiazepines (BDZs), carbamazepine and lithium in dealing with the suicidally inclined.

A. Antidepressants

A major depression would be diagnosable in 40 to 70% of suicidal patients (Barraclough et al. 1972; Montgomery & Montgomery, 1982) and, according to Guze and Robins (1970), 15% of patients with a mood disorder eventually commit suicide. The association between deficient 5-HT transmission and suicidal behavior suggests that administration of the selective serotonin reuptake inhibitors (SSRIs; fluoxetine, fluvoxamine, paroxetine, sertraline, citalopram, zimelidine) should reduce suicidal tendencies. To a certain extent, studies have confirmed this hypothesis: researchers studying zimelidine, citalopram, fluvoxamine (De Wilde, 1985; Gonella et al. 1990; Montgomery et al. 1981; Mullin et al. 1988) have found a significantly greater improvement in suicidal tendencies in the early phase (2 weeks) of antidepressant treatment as compared with amitryptiline, mianserin and dothiepin; in another trial (Muijen et al. 1988), patients on fluoxetine scored significantly lower on suicidal feelings at the end of the study (6 weeks) than patients on mianserin, in spite of similar depression ratings. Consequently, one wonders whether certain ADs have an effect on suicidal tendencies apart from their effect on other aspects of depressive illness. For others, the presence or absence of a history of suicide attempts is a promising guideline in selecting a first-choice antidepressant for a given patient: Sacchetti et al. (1991) found that fluoxetine and clomipramine (both 5-HTergic drugs) were better antidepressants than nortriptyline and desipramine (NEergic drugs) for depressed people with a positive history of suicide attempts. Lopez-Ibor (1993) found that paroxetine was as effective as other antidepressants in rapidly resolving the suicidality associated with depression, and other studies (for review, see Mann et al. 1993) claim an initial advantage, if compared with a reference antidepressant, for the SSRIs in terms of early (up to 4 weeks of treatment) improvement in suicidal ideation. Again, Ottevanger (1993) found that fluvoxamine was more effective than imipramine and placebo in relieving depression in patients with a high initial suicide score. Gasperini et al. (1992), however, did not confirm the findings.

Apart from choosing a medication that is most effective in preventing suicidal behavior, it is important to choose a drug that is less toxic: the newer antidepressants (mianserin, trazodone, nomiphensine and the 5-HTergic class of drugs as a whole) are associated with a higher rate of survival after overdose than TCAs and maprotiline. Among the TCAs, association with overdose lethality is weakest for clomipramine, possibly because it is less cardiotoxic (Schifano et al. 1991). Dothiepin would have a greater toxicity in overdose (Buckley et al. 1994). MAOIs have an intermediate toxicity; moclobemide and venlafaxine seem associated with low toxicity (Feighner 1994). Looking at mianserin (a 2nd-generation AD with NEergic properties), Montgomery and Montgomery (1984) performed a double-blind trial in a high suicidal risk group of personality-disorder patients without depression and found no difference between the active drug's effectiveness in preventing further suicidal attempts and that of a placebo. Given the combination of the apparent superiority of 5-HTergic drugs for treating patients with suicidal behavior and the relatively smaller number of overdose fatalities associated with them, it has been suggested (Baldwin et al. 1991; Maric, 1992; Montgomery, 1990; Preskorn & Burke, 1992; Sacchetti et al. 1991) that these drugs should be prescribed whenever there is a reasonable suspicion of suicide risk in the course of major depression.

However, a report from Teicher et al. (1990) of six depressed patients free of recent serious suicidal ideation who developed intense, obsessive and violent suicidal preoccupation after 2–7 weeks of fluoxetine treatment has recently given rise to lively debate regarding the use of SSRIs in depression (Mann & Kapur, 1991; Power & Cowen, 1992). A few similar case reports involving fluoxetine have since been published (Dasgupta & Hoover, 1990; King et al. 1991; Masand et al. 1991). Paradoxical reactions (patients responding in a manner opposite to clinical expectation, such as increased suicidal tendencies) have been described since the sixties in various case reports of patients receiving antidepressants (including amitriptyline and desipramine). However, only one double-blind study has demonstrated a possible drug-related effect in promoting suicidal behavior. In a double-blind, placebo-controlled, large-sample, year-long trial, Rouillon et al. (1989) found that there was a higher incidence of suicide attempts in the group of patients receiving maprotiline (a NErgic agent) than among those receiving placebo (although maprotiline was clearly superior to placebo in antidepressant efficacy). It may be that increased suicidal tendencies observed in patients administered with ADs reflects lack of change or worsening of depression and/or is secondary to the disorganization of certain vulnerable individuals in response to drug-induced activation and/or akathisia (Hoover, 1991; Lipinski et al. 1989). Another explanation specific to SSRIs has to do with the fact that these drugs, before enhancing 5-HT turnover, cause an initial decrease in the firing rate of 5-HT neurons and, thus, would enhance suicidal tendencies early in the treatment because of an effect on the neurobiologic regulation of the suicide/aggression threshold (Mann & Kapur 1991). Consequently, although it has not been proven that some ADs precipitate suicidal ideation in a small sub-sample of patients, whatever the AD the patient is prescribed, the clinician should assess the severity of depression and suicidal ideation and the occurrence of akathisia (described especially in patients administered with fluoxetine; Wirshing et al. 1992) during treatment.

Anyway, to obtain a relief from depression (in the hope that suicidal behavior will also remit) it is important to follow certain treatment recommendations (for review, see also American Psychiatric Association-1993- and Schifano & De Leo, 1993):

a) dosage of the AD has to be in the correct range: usually 100–300 mg daily for tricyclics such as amitriptyline, imipramine and clomipramine (with a starting dose of 25–50 mg daily); 20–80 mg daily for fluoxetine (with a 20 mg daily starting dose); 50–200 mg daily for sertraline (with a 50–100 mg daily starting dose); 20–40 mg daily for paroxetine (with a 20 mg daily starting dose); 200–300 mg daily for fluvoxamine (with a 50–100 mg daily starting dose);

b) trazodone, bupropion and monoamine-oxidase inhibitors (MAOIs) have been relegated to a 2nd-line position in the treatment of major depression because of antidepressant response and drop-out rate due to side effects; bupropion may be used when the patient has failed sequential trial with tricyclics and SSRIs; MAOIs are to be given when "atypical" symptoms (mood reactivity, irritability, hypersensitivity to rejection, hypersomnia, hyperphagia and psychomotor agitation) characterize the clinical picture (Preskorn & Burke, 1992);

c) patient improvement is seen after 3–6 weeks, so adequacy of response cannot be judged until after this period of time; in the case of a first episode of major depression, the treatment should be continued for at least 14–20 weeks after achieving full remission. Patients who have had multiple episodes of depression should be considered for maintenance treatment. Tapering of the medication (not required for fluoxetine) is not done any faster than 25% of the total dose each week (Thase 1992);

d) initial treatment with ADs fails to achieve a satisfactory response in approximately 30% of patients with a major depressive disorder. In the case of non-response, the diagnosis has to be reviewed and other treatment options considered (simultaneous use of multiple ADs; use of an adjunctive agent; electro-convulsive therapy), but the physician should also take into account a psychiatric consultation/referral.

B. Neuroleptics

Prospective studies have shown that most patients admitted to hospital after attempting suicide are diagnosed as affected by personality disorders (especially borderline, antisocial or histrionic; Montgomery & Montgomery, 1982). Suicidal behavior, hostility and impulsiveness are frequent criteria for the diagnosis of a personality disorder. The mechanisms leading to suicide in borderline patients are different from those of psychiatric patients in general and the absence of depressive symptoms does not mean a lower risk of suicide (Kielsberg, 1991). Both antidepressants (mianserin, as described above; Montgomery & Montgomery, 1984) and neuroleptics have been used in the treatment of such patients. Montgomery et al. (1979) administered flupenthixol depot i.m. or placebo every four weeks to a group of patients with a history of at least three documented suicidal acts. There was a significant reduction in the number of suicidal acts 4 months, 5 months and 6 months after treatment in the group receiving the neuroleptic. What seems a positive effect might be due to a reduction in anxiety level, but if that is the case, it is not clear why mianserin (which also has an anxiolityc effect) did not produce similar results. Although the dependent variable included suicidal attempts (and not completed suicide); the sample size and duration of observation were limited and depot administration could have "biased" the compliance rate, the study is a very important one. The use of NLs has also been recommended in the acute suicidal crisis (Wolfersdorf, 1992).

In relation to the more general problem of the pharmacological treatment of personality disorders, other uncontrolled but long-term studies have suggested the efficacy of low doses of high-strength neuroleptics with "disinhibiting" properties, such as thiotixene, fluphenazine (Brinkley et al. 1979) and trifluperazine (Cowdry & Gardner, 1988). A further reason for the use of neuroleptics in the (indirect) prevention of suicidal behavior could be the high rate (15%) of schizophrenic subjects who successfully commit suicide (Miles, 1977; Pokorny, 1964).

C. Benzodiazepines

GABA-ergic agents such as benzodiazepines could, at least in theory and indirectly, be useful in preventing suicidal behavior, because they reduce anxiety and have a low (but not negligeable; Michel et al. 1994a; Michel et al. 1994b) toxic risk in the case of overdose. However, it has been reported (Gardner & Cowdry, 1985) that alprazolam itself and this class of drugs as a whole provoke a general disinhibition of behavior leading to an increase in self-destructive and impulsive episodes (Gaind & Jacobi, 1978; Salzmann et al. 1974). These drugs are unsuitable for prolonged use, as dependence has been reported after a few months of therapeutic doses (Winokur et al. 1980). Again, Taiminen et al. (1993) found that, at the time of death, suicidal patients, with respect to non suicidal ones, had been receiving a higher dosage of benzodiazepines.

D. Lithium and Carbamazepine

Sacchetti et al. (1991) recently considered whether a history of suicidal attempts affects response to long-term lithium treatment. He found that patients with previous suicidal behavior showed poor lithium stabilization, but they also showed a reduction of over 70% in suicide attempt rates. The constant and attentive treatment of bipolar disorders with lithium may in itself be considered a strategy for preventing suicidal behavior and long-term treatment may lower the risk of suicide (Causemann & Muller-Oerlinghausen, 1988; Coppen et al. 1990; Muller-Oerlinghausen et al. 1988; Volk & Muller-Oerlinghausen, 1988). Extensive use of lithium in recurrent affective disorders has been estimated to prevent as many as 20% of suicides (Barraclough, 1972). It is for this reason that patients are not likely to overdose with lithium (Waddington & McKenzie, 1994). Vetro et al. (1981) also found that lithium treatment improved both outward violence and self-destructiveness in hospitalized hyperaggressive children. Apart from its use in bipolar disorders, lithium is felt by many specialists to be the most effective adjunct to AD treatment: it is reported to be useful in over 50% of patients suffering from major depression who have not achieved a therapeutic response during trial with an AD (American Psychiatric Association, 1993). If it is effective, the drug should be continued for the duration of treatment of the acute episode. Other authors have also suggested that the potential antisuicidal activity of lithium may not necessarily parallel its influence on affective symptoms (Modestin & Schwarzenbach, 1992; Sacchetti, 1991) and is associated with its serotonergic and anti-aggressive properties (Muller-Oerlinghausen et al. 1992). However, the administration of lithium requires a particularly thorough follow-up, good compliance (Di Costanzo & Schifano, 1991) and periodical blood tests, since its margin of safety is narrow.

That is why the use of carbamazepine (Pary et al. 1987) has been suggested instead, as it has a euthymic effect on patients with affective disorders, especially bipolar disorders (Ballenger and Post, 1980; Di Costanzo & Schifano, 1991; Okuma et al. 1981) and it has been reported to reduce the frequency and severity of episodes of behavioral disturbances in various diagnostic subgroups (Gardner & Cowdry, 1986; Mattes, 1984).

IV. Discussion and Conclusions

If the role of psychiatric drugs in the treatment of suicidal patients still remains uncertain, this may be due to methodological issues (experimental design, identification of high-risk groups, evaluation of treatment results, size of the sample) which affect most trials. Nonetheless, a few suggestions seem to emerge.

With regard to the use of ADs, the prevalence of depression in suicidal patients is high enough to warrant appropriate pharmacotherapy (Mendelson et al. 1993). The clinician has to make his own choice of first line treatment. Given the role of 5-HT in promoting suicidal behavior and their low toxicity risk in the event of overdose, the use of SSRIs would seem advisable, also because the balance of evidence does not seem to support any association between these drugs and the emergence of suicidal ideation. On the other hand, Song et al. (1993) emphasize that TCAs are less expensive, equally effective and well tolerated and some claim a superior efficacy of TCAs over SSRIs (Roose et al. 1994). Mianserin and trazodone are well tolerated and safe in overdose, but do they have the same antidepressant effect as TCAs? (Moller et al. 1994). Given the limitations in prescribing MAOIs (due to pharmacological interactions), one could put these drugs in a second line position. Consequently, reversible monoamino-oxidase inhibitors drugs are welcome.

When the clinician has to cope with certain personality disorders, the use of a (depot) neuroleptic for the prevention of suicidal behavior would seem useful, both because of the mode of administration (which ensures compliance) and the reduced behavioral "explosiveness" characterizing these clinical pictures.

With respect to the use of lithium, its beneficial effect on suicidal behavior is now well documented, but its use should be confined to those patients who have had a psychiatric consultation (if not a referral).

Anyway, the physician must remember that the best treatment currently available for suicidal patients comes from understanding the interactive role that diagnosis, medication and psychotherapy play (drugs are not to be prescribed without due consideration of the patient's situation as a whole) (Hendin, 1991). Several issues are to be kept in mind: providing the patient with information about the drugs and their side-effects; making sure that doses are adequate; entrusting the medication to a relative where possible; prescribing drugs only in small quantities to prevent overdose (i.e. no more than a 5-day supply); not permitting refills without writing another prescription (Blumenthal, 1988).

If suicide prevention is successful, this will not generate any statistical data (Murphy, 1984), but the patient will live. On the other hand, as Ennis (1983) points out, everyone who works with suicidal patients has to learn to accept the fact that many patients commit suicide in spite of optimal medical care. When all else fails and a suicide occurs, at least two things remain to be done (Lowenstein, 1985): the physician's first responsibility is to ward the family (survivors often feel guilty) and, later on, the clinician has to ask himself what the most likely causes of suicide were (psychological autopsy), just as he would do for any other patient's death. This will help us all to develop better strategies for suicide prevention.

References

American Psychiatric Association. Practice guideline for major depressive disorder in adults. American Journal of Psychiatry, 1993; 150(suppl.): 1–26.

Asberg M, Traskman M, Thoren P. 5-HIAA in the cerebrospinal fluid. A biochemical suicide predictor? Archives of General Psychiatry, 1976; 33: 1193–1197.

Baldwin D, Bullock T, Montgomery D, Montgomery S. 5-HT reuptake inhibitors, tricyclic antidepressants and suicidal behavior. International Clinical Psychopharmacology, 1991; 6(suppl.): 49–55.

Ballenger NC, Post RM. Carbamazepine (Tegretol) in manic-depressive illness: a new treatment. American Journal of Psychiatry, 1980; 137: 782–790.

Barraclough B. Suicide prevention, recurrent affective disorders and lithium. British Journal of Psychiatry, 1972; 121: 391–392.

Barraclough B, Bunch J, Nelson B. A hundred cases of suicide: clinical aspects. British Journal of Psychiatry, 1974; 125: 355–373.

Blumenthal SJ. Suicide: A guide to risk factors, assessment, and treatment of suicidal patients. Medical Clinics of North America, 1988; 72: 937–962.

Brinkley R, Beitman BD, Friedel RO. Low-dose neuroleptic regimen in the treatment of borderline patients. Archives of General Psychiatry, 1979; 36: 319–326.

Buckley NA, Dawson AH, Whyte IM, Henry DA: Greater toxicity in overdose of dothiepin than of other tricyclic antidepressants. The Lancet, 1994; 343: 159–162.

Cassidy S, Henry J. Fatal toxicity of antidepressant drugs in overdose. British Medical Journal, 1987 295: 1021–1024.

Causemann B, Muller-Oerlinghausen B. Does lithium prevent suicides and suicide attempts? In Birch (editor) Lithium: inorganic pharmacology and psychiatric use, IRL Press, Oxford, 1988.

Cohen LS, Winchel RM, Stanley M. Biological markers of suicide risk and adolescent suicide. Clinical Neuropharmacology, 1988; 11: 423-435.

Coppen A, Standish-Barry H, Bailey J, Houston G, Silcocks C, et al. Long-term lithium and mortality. Lancet, 1990; 335: 1347.

Cowdry RW, Gardner DL. Pharmacotherapy of borderline personality disorder. Archives of General Psychiatry, 1988; 45: 111–119.

Crome P. The toxicity of drugs used for suicide. Acta Psychiatrica Scandinavica suppl, 1993; 371: 33–37.

Dasgupta K, Hoover E. Additional cases of suicidal ideation associated with fluoxetine. American Journal of Psychiatry, 1990; 147: 1570–1571.

De Leo D, Marazziti D. Biological prediction of suicide: the role of serotonin. Crisis, 1988; 9: 109–118.

De Wilde J, Mertens C, Fredricson Overo K, Hepfner Peterson HE. Citalopram versus mianserin. Acta Psychiatrica Scandinavica, 1985; 72: 89–96.

Di Costanzo E, Schifano F. Lithium alone or in combination with carbamazepine for the treatment of rapid-cycling bipolar affective disorder. Acta Psychiatrica Scandinavica, 1991; 83: 456–459.

Ennis J. Suicide prevention. Canadian Medical Association Journal, 1983; 129: 1078.

Feighner JP. The role of venlafaxine in rational antidepressant therapy. Journal of Clinical Psychiatry, 1994; 55(9, suppl A): 62–68.

Gaind R, Jacobi R. Benzodiazepines causing aggression. In Gaind, Hudson (Eds) Current Themes in Psychiatry, Publishing Company, London, 1978.

Gardner DL, Cowdry RW. Alprazolam-induced dyscontrol in borderline personality disorder. American Journal of Psychiatry, 1985; 142: 98–100.

Gardner DL, Cowdry RW. Positive effects of carbamazepine on behavioral dyscontrol in borderline personality disorder. American Journal of Psychiatry, 1986; 143: 519–522.

Gasperini M, Gatti F, Bellini L, Anniverno R, Smeraldi E. Perspectives in clinical psychopharmacology of amitryptyline and fluvoxamine: a double-blind study in depressed inpatients. Neuropsychobiology, 1992; 26: 186–192.

Gonella G, Baignoli G, Ecari V. Fluvoxamine and imipramine in the treatment of depressive patients: a double-blind controlled study. Current Medical Research and Opinion, 1990; 12: 177–184.

Guze SB, Robins E. Suicide and primary affective disorders. British Journal of Psychiatry, 1970; 117: 437–438.

Hendin H. Psychodynamics of suicide, with particular reference to the young. American Journal of Psychiatry, 1991; 148: 1150–1157.

Hoover CE. Suicidal ideation not associated with fluoxetine. American Journal of Psychiatry, 1991; 148: 543.

King RA, Riddle MA, Chappell PB. Emergence of self-destructive phenomena in children and adolescents during fluoxetine treatment. Journal of American Academy of Child and Adolescent Psychiatry, 1991; 30: 179–186.

Kjelsberg E, Eikeseth PH, Dahl AA. Suicide in borderline patients: predictive factors. Acta Psychiatrica Scandinavica, 1991; 84: 283–287.

Lipinski JF, Mallya G, Zimmerman P, Pope HG. Fluoxetine-induced akathisia: clinical and theoretical implications. Journal of Clinical Psychiatry, 1989; 50: 339–342.

Lopez-Ibor JJ jr. Reduced suicidality with paroxetine. European Psychiatry, 1993; 8(suppl.1): 17s–19s.

Lowenstein SR. Suicidal behavior: recognition and intervention. Hospital Practice, 1985; 10: 52–71.

Mann JJ, Goodwin K, O'Brien C, Robinson DS. Suicidal behavior and psychotropic medication. Neuropsychopharmacology, 1993; 8: 177–183.

Mann JJ, Kapur S. The emergence of suicidal ideation and behavior during antidepressant pharmacotherapy. Archives of General Psychiatry, 1991; 48: 1027–1032.

Maric MM. Criteria for selection of first choice antidepressive agents. Ceskoslovenska Psychiatrica, 1992; 88: 220–224.

Masand P, Gupta S, Dewan M. Suicidal ideation related to fluoxetine treatment. New England Journal of Medicine, 1991; 324: 420.

Mattes JA. Carbamazepine for uncontrolled rage outbursts. Lancet, 1984; 2: 1164–1165.

Mendelson WB, Rich CL. Sedatives and suicide: the San Diego study. Acta Psychiatrica Scandinavica, 1993; 88: 337–341.

Michel K, Arestegui G, Spuhler T. Suicide with psychotropic drugs in Switzerland. Pharmacopsychiatry, 1994; 27: 114–118.

Michel K, Waeber W, Valach L, Arestegui G, Spuhler T. A comparison of the drugs taken in fatal and nonfatal self-poisoning. Acta Psychiatrica Scandinavica, 1994; 90: 184–189.

Miles D. Conditions predisposing to suicide: a review. Journal of Nervous and Mental Diseases, 1977; 164: 231 246.

Modestin J, Schwarzenbach F. Effect of psychopharmacotherapy on suicide risk in discharged psychiatric inpatients. Acta Psychiatrica Scandinavica, 1992; 85: 173–175.

Moller HJ, Fuger J, Kasper S. Efficacy of new generation antidepressants: meta-analysis of imipramine-controlled studies. Pharmacopsychiatry, 1994; 27: 215–223.

Montgomery SA. Depression in the elderly: pharmacokinetics of antidepressants and deaths from overdose. International Clinical Psychopharmacology, 1990; 5: 67–74.

Montgomery SA, McAuley R, Rani SJ, Roy D, Montgomery DB. A double-blind comparison of zimelidine and amitriptyline in endogenous depression. Acta Psychiatrica Scandinavica, 1981; 290(suppl.): 314–327.

Montgomery SA, Montgomery DB. Pharmacological prevention of suicidal behavior. Journal of Affective Disorders, 1982; 4: 291–298.

Montgomery SA, Montgomery DB. The prevention of suicidal acts in high risk patients. In Usdin (editor) Frontiers in biochemical and pharmacological research in depression, Raven Press, New York, 1984.

Montgomery SA, Montgomery DB, Rani SJ, Roy DH, Shaw PJ. Maintenance therapy in repeat suicidal behavior: a placebo-controlled trial. 10th International Congress for Suicide Prevention, Ottawa, 1979.

Montgomery DB, Roberts A, Green M, Bullock T, Baldwin D, Montgomery SA. Lack of efficacy of fluoxetine in recurrent brief depression and suicidal attempts. European Archives of Psychiatry and Clinical Neuroscience, 1994; 244: 211–215.

Muijen M, Roy D, Silverstone T, Mehmet A, Christie M. A comparative clinical trial of fluoxetine, mianserin and placebo in depressed outpatients. Acta Psychiatrica Scandinavica, 1988; 78: 384–390.

Muller-Oerlinghausen B, Ahrens B, Grof E, Grof P, Lenz G, et al. The effect of long-term lithium treatment on the mortality of patients with manic-depressive and schizoaffective illness. Acta Psychiatrica Scandinavica, 1992; 86: 218–222.

Muller-Oerlinghausen B, Kossmann B, Volks J, Hermann H. Frequency, quality and temporal pattern of relapses during ten years of lithium prophylaxis. In Birch (editor) Lithium: inorganic pharmacology and psychiatric use, IRL Press, Oxford, 1988.

Mullin JM, Pandita-Gunawardena VR, Whithead AM. A double-blind comparison of fluvoxamine and dothiepin in the treatment of major affective disorders. British Journal of Clinical Practitioner, 1988; 42: 51–55.

Murphy GE. Clinical indication of suicidal risk. Archives of General Psychiatry, 1972; 27: 356–359.

Murphy GE. The prediction of suicide: why is it so difficult? American Journal of Psychotherapy, 1984; 38: 341–349.

Murphy GE. The physician's role in suicide prevention. In Roy (editor) Suicide, Williams & Wilkins, Baltimore, 1986.

Ojehagen A, Regnell G, Traskman-Bendz L. Deliberate self-poisoning: repeaters and non-repeaters admitted to an intensive care unit. Acta Psychiatrica Scandinavica, 1991; 84: 266–271.

Okuma T, Inanaga K, Otsuki S. A preliminary double-blind study of the efficacy of carbamazepine in prophilaxis of manic-depressive illness. Psychopharmacology, 1981; 73: 95–96.

Ottevanger EA. Fluvoxamine superior to imipramine in suicidal depressed patients. European Neuropsychopharmacology, 1993; 3(3, special issue): 362–363.

Pary R, Lippmann S, Turns DM, Tobias CR. Drug selection after overdose recovery: carbamazepine or lithium. Journal of Kentucky Medical Association, 1987; 85: 21–23.

Pokorny AD. Suicide rates in various psychiatric disorders. Journal of Nervous and Mental Diseases, 1964; 139: 499–506.

Power AC, Cowen PJ. Fluoxetine and suicidal behaviour: some clinical and theoretical aspects of a controversy. British Journal of Psychiatry, 1992; 161: 735–741.

Preskorn SH, Burke M. Somatic therapy for major depressive disorder: selection of an antidepressant. Journal of Clinical Psychiatry, 1992; 53(suppl.): 5–16.

Rao ML, Braunig P, Papassotiropoulos A. Autoaggressive behavior is closely related to serotonin availability in schizoaffective disorder. Pharmacopsychiatry, 1994; 27: 202–206.

Robins E, Murphy GE, Wilkinson RM. Some clinical considerations in the prevention of suicide based on a study of 134 successful suicides. American Journal of Public Health, 1959; 49: 888–898.

Roose SP, Glassman AH, Attia E, Woodring S. Comparative efficacy of selective serotonin reuptake inhibitors and tricyclics in the treatment of melancholia. American Journal of Psychiatry, 1994; 151: 1735–1739.

Rouillon F, Phillip R, Serrurier D, Ansart E, Gerard MJ. Rechutes de depression unipolaire et efficacite de la maprotiline. Encephale, 1989; 15: 527–534.

Sacchetti E, Vita A, Guarneri L, Cornacchia M. The effectiveness of fluoxetine, clomipramine, nortriptyline and desipramine in major depressives with suicidal behaviour: preliminary findings. In Cassano, Akiskal (editors) Serotonin-related psychiatric syndromes: clinical and therapeutic links, No. 165, pp. 47–53, Royal Society of Medicine Services Limited, London, 1991.

Salzmann C, Kochansky GE, Shader RI, Porrino LJ, Harmatz JS, et al. Chlordiazepoxide-induced hostility in a small group setting. Archives of General Psychiatry, 1974; 31: 401–405.

Schifano F, De Leo D: Can pharmacological intervention aid in the prevention of suicidal behavior? Pharmacopsychiatry, 1991; 24: 113–117.

Schifano F, Magni G, Corfini A. Antidepressivi e apparato cardiovascolare. Aspetti farmacologici e clinici. Quaderni Italiani di Psichiatria, 1991; 10: 221–244.

Schifano F. Pharmacological strategies for preventing suicidal behaviour. CNS Drugs, 1994; 1: 16–25.

Song F, Freemantle N, Sheldon TA, House A, Watson P, et al. Selective serotonin reuptake inhibitors: meta-analysis of efficacy and acceptability. British Medical Journal, 1993; 306: 683–686.

Taiminen TJ. Effect of psychopharmacotherapy on suicide risk in psychiatric inpatients. Acta Psychiatrica Scandinavica, 1993; 87: 45–47.

Thase ME. Long-term treatment of recurrent depressive disorders. Journal of Clinical Psychiatry, 1992; 53 (suppl.): 32–41.

Teicher MH, Glod C, Cole JO. Emergence of intense suicidal preoccupation during fluoxetine treatment. American Journal of Psychiatry, 1990; 147: 207–210.

Traskman L, Asberg M, Bertilsson L, Sjostrand L. Monoamine metabolites in cerebrospinal fluid and suicidal behavior. Archives of General Psychiatry, 1981; 38: 631–636.

Van Praag HM. Psychobiology of suicidal behavior. Annals of New York Academy of Sciences, 1986; 487: 150–167.

Vetro A, Pallag P, Szentistvanyl LI, Vargha M, Szilard J. Treatment of childhood aggressivity with lithium. Aggressologie, 1981; 22: 27–30.

Volk J, Muller-Oerlinghausen B. Quality of interepisodic periods in patients with affective disorders under long-term lithium treatment. Pharmacopsychiatry, 1988; 21: 426–427.

Waddington D, McKenzie IP. Overdose rates in lithium-treated versus antidepressant-treated outpatients. Acta Psychiatrica Scandinavica, 1994; 90: 50–52.

Winokur A, Rickels K, Greenblatt DJ, Snyder PJ, Schatz NJ. Withdrawal reaction from long-term, low dosage administration of diazepam. Archives of General Psychiatry, 1980; 37: 101–105.

Wirshing W, Van Putten T, Rosenberg J, Marder S, Ames D, et al. Fluoxetine, akathisia, and suicidality: is there a causal connection? Archives of General Psychiatry, 1992; 49: 580–581.

Wolfersdorf M. The value of psychotropic drugs in treatment of suicidal behavior. Psychiatrische Praxis, 1992; 19: 100–107.

22. Risk Factors for Non-Compliance with Outpatient Aftercare

Implications for the Management of Attempted Suicide Patients

C. VAN HEERINGEN, C. JANNES, W. BUYLAERT and H. HENDERICK

I. Introduction

General agreement exists with respect to the need for treatment of underlying psychiatric or psychological problems in the majority of suicide attempters after emergency assessment and crisis intervention. In a growing number of studies the efficacy of such psychotherapeutical or psychopharmacological treatments is addressed. A few studies, however, have addressed a topic that can be considered as very important with respect to the evaluation of treatment of suicide attempters, i.e. the extent to which attempters actually go into treatment. Available data indicate that less than 50% of attempted suicide patients comply with referral to outpatient aftercare (Kurz & Moller, 1984). Despite these findings, research on this problem among attempted suicide patients has been rather scarse. High repetition rates with strongly increased risks of completed suicide following attempted suicide, however, indicate the necessity of such research.

Outside the suicidological area the problem of non-compliance has been studied more extensively. The most common risk factors for non-compliance can be categorized to four groups (Nivel, 1989; Van Heeringen, 1992). A first category includes risk factors related to the patients, and more specifically their attitudes and illness behaviour, and to the presence of social supervision (i.e. control of compliance by partner, family or others). A second category includes illness-related factors (like duration and nature). The third category consists of behaviours, attitudes and cognitions of treating therapists, such as their belief in the prognosis of the illness. A final category includes characteristics of the treatment setting such as the presence or absence of continuity of care.

D. De Leo, A. Schmidtke and R.F.W. Diekstra (eds.), Suicide Prevention, 211–218.

The available literature indicates that compliance is the result of a complex interaction between these risk factors. Within the context of suicidology, research has focused mainly on the first and last category of risk factors (Van Heeringen, 1992). For instance, no association between compliance and sociodemographic or clinical characteristics (O'Brien et al. 1987) or motivation for treatment (Torhorst et al. 1988) has been found. Regarding the fourth category (characteristics related to the treatment setting) the arrangement of a fixed appointment (Müller & Geiger, 1981) and continuity of therapist (Torhorst et al. 1988) were found associated with compliance.

These studies provide an insight in the risk factors for non-compliance with outpatient aftercare among attempted suicide patients following treatment in the hospital. However, as no studies have addressed the complex interactive pattern of these risk factors, the present study aimed at the assessment of the independent effect of potential risk factors on compliance by means of multivariate analyses.

II. Methods

Patients included in the study were suicide attempters who were consecutively referred to the Accident & Emergency Department of the University Hospital Gent between January 1st 1987 and December 31st, 1990. Attempted suicide was defined according to Hawton & Catalan (1987), and included deliberate self-poisoning ("the deliberate ingestion of more than the prescribed amount of medical substances, or ingestion of substances never intended for human consumption, irrespective of whether harm was intended") and deliberate self-injury ("any intentional self-inflicted injury, irrespective of the apparent purpose of the act").

All patients were examined by a resident in training. Sociodemographic and clinical characteristics, and characteristics of the referral procedures were monitored using a standardized monitoring form. A test-retest study indicated a good to excellent interassessor-reliability of sociodemographic characteristics and of the assignment of psychiatric diagnoses (Van Heeringen et al. 1993).

Criteria for inclusion in the study were: age 15 years or older, living in the city of Gent and its suburbs (thus reducing interference with compliance for geographical reasons), and referral to the Outpatient University Department of Psychiatry or to a Community Mental Health Service following discharge from the Accident & Emergency Department or from the Inpatient University Department of Psychiatry. Referral occurs preferably with a fixed appointment, which was arranged by the resident who performed the patient interviews; in case no fixed appointment can be provided (when patients are discharged at night or during the weekend), patients are given written information on an appropriate outpatient facility, and adviced to contact this facility within one week following discharge.

Compliance was considered present if the outpatient (mental) health facility or professional was attended at the moment of the fixed appointment, or in case of referral without a fixed appointment, within a week after discharge from the hospital. Compliance was assessed by contacting the involved outpatient facility two weeks after the fixed appointment, or two weeks after the patient was given advice to contact this facility.

III. Results

During the study period 378 patients met the criteria for inclusion. As shown in Table 1, the majority of these patients were female, aged 30 years or older, not living alone, unmarried or divorced, and employed or economically inactive. With respect to psychiatric diagnoses, adjustment disorder, substance abuse disorder, and affective disorder were among the most common DSM-III-R axis-I diagnoses. A formal axis-II disorder was diagnosed in approximately 1 in 4 patients. The suicide attempts were most commonly characterized by deliberate self-poisoning, with a low risk of repetition as assessed by means of the Buglass & Horton Prediction Repetition Scale (Buglass & Horton, 1974). Fifteen percent of the patients was in outpatient treatment at the moment of the suicide attempt. Approximately 28% of the patients reported a history of outpatient treatment, or previous suicide attempts. Following treatment in the Accident & Emergency Department about one third was admitted to the Inpatient Psychiatric Department. In view of outpatient aftercare, the majority were referred to the Outpatient Psychiatric Department of the University Hospital (OPD). In the OPD treatment was offered by another psychiatrist than the psychiatrist who interviewed the patients at the Accident & Emergency department and who arranged the referral. The arrangement of a fixed appointment was included in the referral-procedure in 44%.

The next step in the analysis of the data was the calculation of the strength of the associations between these characteristics and the occurrence of non-compliance with referral to outpatient aftercare. Table 2 gives an overview of the results of the univariate logistic regression analyses. Six of the above mentioned characteristics were significantly associated with non-compliance. The strongest identified risk factor was the absence of a fixed appointment: when patients were referred without an appointment the risk of non-compliance was increased more than threefold (OR=3.53) when compared to patients referred with an appointment. Similarly, increased risks were found if patients were no outpatient at the moment of the attempt (OR=3.21) or before the attempt (OR=2.30), and if patients showed a substance abuse disorder (OR=2.30). Inpatient psychiatric treatment after discharge from the Accident & Emergency Department was associated with a lower risk of non-compliance than discharge to home following treatment in the Accident & Emergency Department (OR=1.69).

TABLE 1

Characteristics of study population

Characteristic	n	%
Sex		
male	151	39.9
female	227	60.1
Age (years)		
<30	159	42.1
>30	219	57.9
Situation of living		
alone	81	21.4
not alone	297	78.5
Educational level		
primary	107	28.3
higher	271	71.7
Marital status		
married/widowed	180	47.6
unmarried/divorced	198	52.4
Employment status		
employed/inactive	314	83.1
unemployed	64	16.9
Psychiatric diagnosis		
substance abuse disorder	59	15.6
mood disorder	59	15.6
adjustment disorder	178	47.1
personality disorder	93	24.6
Method of suicide attempt		
deliberate self-poisoning	345	91.3
deliberate self-injury	33	8.7
Risk of repetition		
low	353	93.4
moderate or high	25	6.6
Psychiatric treatment at the moment of suicide attempt		
outpatient	56	14.8
no outpatient	322	85.2
History of psychiatric treatment		
outpatient	103	27.2
no outpatient	275	72.8

Table 1 (continued)

Characteristic	n	%
History of suicide attempts		
present	108	28.6
Discharge following emergency treatment		
to home	242	64.0
to inpatient psychiatry	136	36.0
Fixed appointment for referral		
present	166	43.9
Referral to		
Community Mental Health Service	163	43.1
Outpatient Psychiatric Dept	215	56.9
All	378	100,0

Finally, it appeared that referral to a Community Mental Health Service (CMHS) was associated with a lower risk of non-compliance than referral to the Outpatient Psychiatric Department of the University Hospital (OR=0.44).

Further analysis pursued the calculation of the independent effect of the above mentioned variables. Multivariate logistic regression analysis using a model containing all variables showed that three characteristics remained significantly associated with non-compliance. These characteristics included discharge from the Accident & Emergency Department to home when compared to referral for inpatient psychiatric treatment (OR=1.77), and referral without a fixed appointment (OR=8.16). Finally, the risk of non-compliance associated with referral to a CMHS following inpatient treatment was found significantly lower when compared to referral to the Outpatient Psychiatric Department (OR=0.21).

IV. Discussion

This study clearly demonstrated, first, that non-compliance with referral to outpatient aftercare constitutes a major problem in the management of attempted suicide patients, as less than half of the patients complied with referral. This finding is in keeping with the results from previous studies (Van Heeringen, 1992). Therefore, studies evaluating the treatment of attempted suicide patients should be interpreted cautiously, as only a minority of patients actually take up treatment. In a controlled intervention study we could recently demonstrate a significant beneficial effect of home visits by a

TABLE 2

Potential risk factors for non-compliance with outpatient aftercare, with relative odds estimates (OR) based on univariable unconditional logistic regression, and 95% Confidence Intervals (CI)

Potential risk factor	Non compliance n	%	Reference category	OR	95% CI	χ^2 value
Sociodemographic characteristics						
male	71	47.0	female	1.05	0.81–1.36	0.159
age < 30 yrs.	72	45.3	> 30 yrs.	0.97	0.64–1.46	0.026
living alone	38	46.9	not living alone	1.07	0.66–1.76	0.064
low education	56	52.3	higher education	1.47	0.94–2.31	2.993
unmarried/divorced	88	44.4	married/widowed	0.89	0.57–1.40	0.293
unemployed	29	47. 5	employed/inactive	1.11	0.52–1.56	0.143
Psychiatric diagnosis						
mood disorder	23	39.0	no mood disorder	0.72	0.41–1.27	1.296
adjustment disorder	82	46.7	no adjustment dis.	1.02	0.68–1.53	0.012
substance abuse dis.	24	63.2	no substance abuse	2.30	1.21–4.64	5.739**
personality disorder	41	44.1	no personality dis.	0.91	0.57–1.46	0.140
Attempt characteristics						
self-poisoning	159	46.1	self-injury	1.16	0.56–2.39	0.163
high risk repetition	13	43.3	low risk	1.34	0.49–3.65	0.331
History of suicide attempts or psychiatric treatment						
previous attempt	49	45.4	no previous attempt	0.99	0.63–1.56	0.001
no previous outpatient	32	31.1	previous outpt.	2.30	1.42–3.72	12.139 ***
Treatment at time of attempt						
no outpatient	13	23.2	outpatient	3.21	1.66–6.19	13.752***
Discharge and referral characteristics						
discharge to home	122	50.4	to psychiatry	1.69	1.10–2.60	5.849**
referral to CMHS [1]	56	34.4	to outpatient dept	0.44	0.29–0.67	15.035***
without appointment	125	59.0	with appointment	3.53	2.29–5.45	33.863***
All	173	45.8				

[1] CMHS = Community Mental Health Service

** P < 0.01; *** P < 0.001

community nurse on compliance with outpatient aftercare among attempted suicide patients (Van Heeringen et al. in press).

Secondly, the present study shows that compliance is not associated with sociodemographic or clinical characteristics (with the exception of substance abuse disorder in univariate analysis). Similarly, no association is found between compliance and characteristics of the suicide attempts. These find-

TABLE 3

Factors associated with non-compliance with outpatient aftercare: results from multivariate logistic regression analysis

Characteristic	OR	95% CI	Reference group	P
discharge to home	1.77	1.06–2.96	to psychiatry	0.03
referral to CMHS [1]	0.21	0.12–0.36	to outpatient dept	0.0000
without appointment	8.16	4.56–14.61	with appointment	0.0000

[1] CMHS = Community Mental Health Service

ings are in keeping with the results from previous studies (Van Heeringen, 1992).

However, previous or current contacts with outpatient psychiatric treatment facilities, or inpatient psychiatric treatment before referral to outpatient aftercare are associated with a significantly reduced risk of non-compliance. Unfortunately, the available data do not allow to distinguish between causality or a selection phenomenon as possible interpretations of this association. Causality would indicate a causal association between psychiatric treatment and subsequent compliance with outpatient aftercare. This means that previous outpatient or inpatient treatment actually leads to increased compliance with subsequent outpatient aftercare. This finding can, however, also be explained as the consequence of a selection bias, indicating that both previous psychiatric treatment and compliance with referral to outpatient aftercare are associated with a common third factor, such as motivation for treatment.

Torhorst et al. (1988) demonstrated, however, that motivation for treatment is not associated with compliance.

The results of this study show that referral to a Community Mental Health Service (CMHS) is associated with a lower risk of non-compliance than referral to the Outpatient Psychiatric Department of the University Hospital (OPD). This finding is more difficult to interpret. The choice between CMHS or OPD was based mainly on geographical grounds. Thus, the results indicate that aftercare is preferably provided close to the patient's homes. However, the concept of continuity of care has attracted some attention in compliance-related research, and evidence was found for a beneficial effect of continuity of therapist (Torhorst et al. 1988). The finding of the present study suggests that continuity of institution (i.e. referral to outpatient aftercare in the same institution), at the least, does not contribute to the enhancement of compliance.

Finally, the absence of a fixed appointment strongly increases the risk of non-compliance. This result provides additional evidence for the finding of Moller & Geiger (1981), who demonstrated in their study an increase of compliance from 31% to 55% by means of the arrangement of a fixed appointment.

The results of the present study have considerable implications for the management of attempted suicide patients. Patients discharged to home following emergency treatment should be considered at high risk of non-compliance. All patients, but especially those at high risk of non-compliance, should be discharged from the hospital with a fixed appointment for outpatient treatment in a facility that is close to their homes from a geographical point of view.

Acknowledgements

This study was supported by a grant from the Belgian National Fund for Scientific Research (NFWO, grant nr. 3.0061.86). The authors also wish to thank the following mental health workers from the participating Community Mental Health Services, for their continued interest and willingness to collaborate: A. Couck (Guidance Centrum), E. Herman (De Schelp), A. Reyniers (Centrum voor Geestelijke Gezondheidszorg, Fabiolalaan), F. Temmerman (CAT), and K. Van Kerckhove (ABC). Dirk De Bacquer is thanked for statistical advice.

References

Buglass D, Horton J. A scale for predicting subsequent suicidal behaviour. Br J Psychiatry, 1974; 124: 573–578.

Hawton K, Catalan J. Attempted suicide: a practical guide to its nature and management. Oxford: Oxford University Press, 1987.

Kurz A, Moller HJ. Hilfesuchverhalten und Compliance von Suizidgefährdeten. Psychiatr Praxis, 1984; 11: 6–13.

Moller HJ, Geiger V. Möglichkeiten zur "Compliance" Verbesserung bei Parasuizidenten. Crisis; 2: 122–129.

Nivel. Patient compliance: a survey of reviews (1979–1989). Netherlands Institute of Primary Health Care, 1989; 21.

O'Brien G, Holton AR, Hurren K, Watt L, Hassanyeh F. Deliberate self-harm and predictors of outpatient attendance. Br J Psychiatry, 1987; 150: 246–247.

Torhorst A, Burk F, Kurz A, Wachtler C, Möller HJ. Motivation for compliance with outpatient treatment of patients hospitalized after parasuicide. In: Moller HJ, Schmidtke A, Welz R (editors). Current issues of suicidology. Berlin: Springer Verlag, 1988.

Van Heeringen C. The management of non-compliance with outpatient aftercare in suicide attempters: a review. It J Suicidol, 1992; 2: 79–83.

Van Heeringen C, Jannes S, Buylaert W, Henderick H, De Bacquer D, Van Remoortel J. The management of non-compliance with referral to outpatient aftercare among attempted suicide patients: a controlled intervention study. Psychol Med (in press).

Van Heeringen C, Rijckebusch W, De Schinkel K, Jannes C. The reliability of the assessment of suicide attempters. Arch Public Health, 1993; 51: 443–456.

23. Implementation of the Suicide Prevention Strategy in Finland

First Follow-up

MAILA UPANNE

The challenge of the implementation of the national suicide prevention project in Finland (1992–96) is to produce focused, practical and effective preventive activities in the country according to the lines adopted in the national strategy. Implementation is the third stage of the national project utilizing the second stage: strategy of suicide prevention published in 1992. The strategy was developed from empirical results and expert recommendations evolved in the first stage: one year's research phase in 1987.

Activities are running on two levels: on local level as spontaneous activities in different fields and on national level as developing and coordinating activities of the project team. The project is run in cooperation with the ministry of social affairs and health.

The intermediate targets and criteria for the national implementation strategy are

- comprehensiveness; taking responsibility all over the country
- multisectorality; different fields co-operating
- multiprofessionality
- involving public services and normal contexts
- utilizing networking structure of committed professionals and
- developing method of cooperation and interactive planning with the sectors.

I. A Nationwide Start

The nation-wide launching of the project was evaluated by means of an extensive (N=6000) inquiry in 1993. The survey was at the same time an intervention; it reminded people of the project and challenged them to take part. Testing the suitability of a questionnaire for evaluation was a part of a task.

1800 (30%) replies was a good field response considering a multiple coverage in organizations. The data may be regarded as describing the state of the

D. De Leo, A. Schmidtke and R.F.W. Diekstra (eds.), Suicide Prevention, 219–223.

work in the beginning of the project rather well; the replies came evidently from units "who had something to tell". One positive phenomenon is worth while mentioning: 40% of the respondents were men.

The data demonstrate that suicide prevention work started well and on a wide scale throughout Finland in 1992. The aim of generating responsibility for preventing suicides was widely accepted. Essential point is, that data concern independent actors and spontaneous activities, not for example passive targets of a campaign.

II. In Survey

About 1200 (66%) respondents mentioned some activity or expressed an active interest in some other way. Some 650 (11%) mentioned to have begun organizing activities. Development projects and training programmes both numbered over 200 (4%).

Close on 1100 (18%) professionals in different fields wished to be included in the contact network as the agreed representatives of their units. A start had been made to planning operations. 95 work units (6%) already had their own planning team and 189 (11%) had appointed a person in charge. Activities emerged all over the country; in all twelve provinces.

Although there were projects under way throughout the country and in all sectors, units engaging in organized, diverse action were still rather few in number. In the whole of Finland only 88 of the units (5% of all respondents) could be classified as "development centres" with organized activity, projects and training. The units carrying out one of these three forms of work numbered 737 (42%).

Suicide prevention was not supported by administrative decisions, plans or funds. It was adopted as an administrative goal only in 300 work units (17%). The work consisted mainly of professional and administrative developing the contents of work; 80–90% of activities was included in the worker's or team's own, normal, local work.

III. Multisectorality

The project aims to develop suicide prevention as a multi-target, multi-sector activity in which various triggering, predisposing and protective mechanisms receive the right kind of attention. The work is motivated by the objective by which each field of expertise makes effective use of its own special know-how and perspective.

The replies indicate that this interim goal has been well achieved even at the initial stage of the project. The inquiry yielded reports from all the core sectors of society. The contact network also represents eleven sectors.

Although activities were being organized in all sectors, most progress in the development of activities (43% of the activities and 66% of the multiple activities) had been made by health care.

The inquiry proved that multi-sector and multi-professional cooperation has made its breakthrough as an operational strategy. For example, at least four different operating sectors were in most cases represented in the planning teams appointed by work units. Regional cooperation is also functioning. One respondent in four reported participation in some joint team involving various administrations. This means some 1700 professionals from 12 fields working in local multisectoral teams in the year of the inquiry. Most (2/3) of the 220 development projects were joint ventures involving two or more administrations, though the leader was in most cases (70%) health care. One third of the projects involved representatives of at least four different organizations. Furthermore, some 300 respondents (16%) were members of a regional crisis team.

So the principle of multi-professionalism seems to have been realized. The five biggest active professional groups, almost equal in size – psychologists, doctors, nursing staff, church workers (clergymen and deacons) and social workers – accounted for two thirds of the respondents and were key groups as developers, as persons in charge and in the network.

IV. Developmental Targets

The targets represent many kinds of practical activities the main aim being developing models for activities. The stage of the process was reflected by many preparatory activities like planning and education.

A considerable proportion (roughly 2/3) of the 220 developing projects aimed to enhance the expertise of the professional personnel and to develop modes of operation. One third of the projects concerned the population direct. Preliminary classification using the project strategy model tentatively revealed that the practices of suicide prevention aim as a whole at a broad, multifocused and multilevel prevention strategy. It can be conceived of as a complex policy affecting various stages in the life process, as is stated in the project strategy.

A large proportion of the projects (42%) dealt with specific prevention; action directed at acute risk of suicide. Many of the projects in this category were, however, in the nature of an initial orientation at the general level when the objective was only just being defined. Unspecific prevention, such as the question of depression, did not yet feature in these projects. Yet over a quarter of the projects dealt with the handling and treatment of crisis situations. This also included postvention; support for the survivors (6%). There were also some promotive projects aimed at helping people to cope with life (7%). In many cases (36%) the projects proceeded by integrating many objectives and modes of operation. The development of practical modes of operation, in most cases for use in treatment, was considered to be the primary functional goal (33%). One project in ten aimed to create and organize the prerequisites for development work. Some projects had a training, research or communication orientation. The progress of the projects was regarded as satisfactory; 61% stated that the goals had been achieved at least reasonably well.

V. Training

It seems, that professional training is regarded one of the most important starting points with a view to development. Over 200 of the respondent agents had arranged training during the year of the inquiry. The majority of the training events had been short, lasting at most one working day (63%). These events made up a total of 2320 hours, which in terms of a seven-hour working day represents 331 working days or a training spell of almost a year. Almost every day there were some 50 professionals receiving training on the subject.

VI. 30 000 Professionals and Half a Million Others

The respondents' estimate of the number of people reached in their work provides an encouraging picture of the way operations have been set in motion. Around half a million members of the public at large was estimated to have fallen within the projects' domain (including public education). The projects are estimated as having involved about 12 000 professionals. In addition to this, about 17 000 workers took part in professional training. This means roughly 30 000 professionals involved in some way or another.

VII. Can We Do Anything?

The majority of the respondents have a positive view of their own ability and that of their unit to act. The replies are encouraging; the respondents appear to have established their own responsibility. As many as 82% of the respondents reckoned their unit could do at least something, and close on one third that it could do a lot. The clergymen, psychologists and teachers at social welfare colleges were most optimistic. Only just one fifth doubted their ability. The most doubtful (who felt there was little or nothing they could do) were people in the fire and rescue service (47%) and doctors (20%).

VIII. The National Project Team

On national level the project has been working – in addition to activating and follow-up activities – with some twenty subprojects developing practical models for suicide prevention.

Topics concern

- models for good care for suicide attempters and people with alcohol problems,
- models for coping with crisis situations for schools, for conscripts, for the police, for those indebted,
- models for regional planning of prevention,
- models for public information,
- strategy for national implementation including a network of 1200 members and for ex. a four times a year-paper.

Projects are run in cooperation with different sectors involved like school, Armed Forces, church, police, hospital districts and health centers. In addition to these there are other sectors involved like provincial administration, some communes, association for guidance for those indebted, professional union of journalists and other associations.

A separate project for prevention and care of depression has been started recently as a part of the Suicide prevention project.

IX. Firm Foundations

The project has got off to a good start, has generated broad expert responsibility, and the first interim goals, such as nation-wide, multi-sector and multi-professional coverage and broad strategy have been achieved. Suicide prevention has begun at practical level. Specially the method of interactive cooperation with other sectors has given a good feedback. During the operational evaluation of the project in 1995–1996 data will be collected on work carried out in Finland as a whole and on the subprojects executed by the project team. The evaluation of the project as a whole will consist of the project's own assessment of operations (process evaluation) and evaluation of the effectiveness of the project and of external national and international evaluation.

The trend in suicide statistics, as many other problem indicators in Finland, has been declining since 1990 (30.3) being 27.6 per 100 000 in 1993.

24. Venlång: The Swedish National Programme for Suicide Prevention

JAN BESKOW and DANUTA WASSERMAN

The scientific background of the Swedish National Programme for Suicide Prevention is the same as in other such programmes. If it in any way distinguishes itself it may be in the ambition to formulate a philosophy for suicide prevention attuned to the present cultural situation regarding suicide. The latter is characterized by severe tensions between divergent attitudes, in which suicide may be perceived as a human right while at the same time a tabooed and despicable act.

For a healthy individual the thought of suicide may generate feelings of freedom and control, very different from the feelings of the severely stressed and depressed suicidal person, who desperately seeks attention and help. The most decisive components of suicide prevention emerge in meetings between such healthy and unhealthy persons. It is therefore the task for a National Programme for Suicide Prevention to present notions and values that may stimulate such encounters.

History

The number of murder and manslaughter cases per 100 000 inhabitants in the city of Stockholm was very high in the 16th century but gradually decreased in the middle of the 18th century, when it reached about the same level as today (Söderberg, 1993). This occurred before the development of the modern state governed by law with a strong police corps at its disposal. This decrease in violence may rather be seen as part of a continually developing civilization process. With these historic figures in mind, today's increasing rates of violence are only ripples on the surface.

On the contrary suicides were relatively rare before the middle of the 18th century but then increased up to the current level. During the 16th and early 17th century the curious but quantitatively not unimportant phenomenon of *suicidal homicides* appeared in Stockholm (Jansson, 1994). Typically a lonely and depressed woman murdered a child and then turned herself in to the police in order to be convicted to death. The gains were threefold. The murdered

D. De Leo, A. Schmidtke and R.F.W. Diekstra (eds.), Suicide Prevention, 225–234.

child was innocent and would be accepted in paradise. The female murderer got absolution for her sins before execution and could thus reach the same paradise, which would have been impossible if she had committed suicide. The primary gain, however, was that she escaped an intolerable life.

The phenomenon of suicidal homicide is understandable in light of the strong religious and judicial sanctions against suicide that prevailed for more than 1500 years. The presuppositions for these homicides weakened at the end of the 17th century. More liberal attitudes and milder penalties made capital punishment less probable. Also increasing tolerance towards suicidal behaviour provided new options and suicidal homicides were no longer the best alternative. This increasing tolerance led in Sweden to a decriminalization of suicidal acts in 1864. The last religious restrictions were withdrawn in 1909.

With this history in mind we must perceive the control of outward aggression as the first step in the civilization process of violence. The next challenge is to be able to reduce suicidal behaviour.

The Present

Cultures build up strong systems of ideas and rites as protection against the fear of death. During the development of the individualized and secular western society traditional cultural and religious concepts about death and suicide have weakened. Human contacts protected by belief systems and rites were partly replaced by the defense mechanism of isolating dying persons in hospitals. Personal experiences of the dying became therefore less frequent and were insufficiently worked through. Urbanization also rendered experiences of dying animals less common.

After World War II with concentration camps and nuclear bombs, followed by further wars, starvation and environmental hazards, the denial of death was totally impossible. Technology, media and population movements made the globe smaller, making way for rapid changes with further individualization, pluralism and ideological uncertainty. As a further stress television and other media supply us with a never-ending flood of suffering and death, which the individual can neither influence nor avoid.

The human rights movement appeared as a rational protection against the cruelties of World War II. Some people also argue that suicide should be seen as a human right, leading to obligations for society and especially physicians in helping people to realize this right. Although this attitude has been formally accepted neither in Sweden nor by international organizations it has influenced the way people think about these matters. An opposing position is to stress the right to one's own *life,* which may lead to active work within the international movement for suicide prevention.

These factors led to complicated attitudes towards suicide and also towards suicide prevention. Suicidal acts are evaluated both as miserable and admirable, as a desperate cry for help and as a philosophical act. The suicidal person is seen both as a coward and as a brave man. Suicide prevention may be regarded as a spontaneous reaction to an urgent need for help and as a violation of personal integrity. The belief systems may be weak both in the suicidal person and in his potential rescuer. Suicide prevention must take place in this domain of great tensions.

Scanty Language

A consequence of the long period of repression is the underdevelopment of the language available to describe suicidal phenomena (Pabst Battin, 1982). Only one word is necessary if a phenomenon is totally forbidden. On the other hand, many different words are required if a problem is to be treated in a differentiated fashion.

In the area of suicidal acts very different concepts have been denoted by the same term. Both a fanatic Buddhist monk's attempt to commit suicide in order to exert political pressure and a young Swedish girl's attempt to escape a troublesome situation for a while by taking a double dose of sleeping pills are referred to as "suicide attempts". In contrast we have many terms for describing homicide, such as murder of 1st degree, manslaughter, maltreatment leading to death, accident, death penalty, war acts etc. Every one of these is associated with different notions and modes of action.

Suicide has different meanings for an individual far from a suicidal act compared to a near-suicidal person. Usually the healthy person does not contemplate suicide. If he does think about death and suicide it is a way of asking sincere questions about the meaning of life and thus a normal part of the life struggle. "Suicide" is then often associated with freedom, control and the possibility to escape heavy and unworthy burdens. Suicide is one of several possibilities.

The situation contrasts sharply to that of the despaired and depressed person for whom suicide may be the unwanted but only possible solution in a situation of intolerable psychic or somatic pain. Such a person may never cease to hope that someone may observe his predicament and offer a better alternative. The problem is that the potential rescuer is the healthy person with *his* attitudes to suicide.

As a consequence of the long history of taboo suicide is often perceived as a very special but isolated problem. This may be true both for the healthy person's less complicated attitude and for the more compulsive experience of the suicidal person. Both may have difficulties in understanding the relations between background factors such as mental disorder, abuse, interpersonal problems and accumulated life problems, and the suicidal problems.

Previous Development of Suicide Prevention

Two thousand people die from suicide and some 20 000 people make suicide attempts each year in Sweden. Since 1970 a slightly declining trend in suicide mortality has been observed, more evident among men than among women. Decreasing rates of death due to infectious diseases and accidents have rendered suicide relatively more important as a cause of death.

Research in suicidology has developed during the last decades in the psychiatric research departments in Stockholm (Karolinska Hospital, head: professor Marie Asberg) and Lund (head: professor Lil TraskmanBendz), concentrating on the treatment of patients with depression and/or suicidal behaviour. A research group in sociomedical suicidology in Goteborg (head: associate professor Jan Beskow), has investigated different groups of suicide using the psychological autopsy method. Specialist training courses in suicidology have been held since 1976. Courses have also been offered to mental health personnel.

A National Centre for Suicide Research and Prevention has been set up at the National Institute for Psychosocial Factors and Health, Karolinska Institute in Stockholm, following a decision by the Swedish parliament to meet the national need for knowledge development in the field of suicide prevention. This national centre is integrated with Stockholm County Council Centre for Suicide Research and Prevention at Karolinska Hospital that has been operating since 1993 (head: professor Danuta Wasserman).

The centre is responsible for the study of attempted suicide in Huddinge, which along with the Psychiatric department at Umea University (head: professor Lars Jacobsson) are the two Swedish settings in the WHO/EURO multicentre study of attempted suicide.

As early as 1983 the National Board of Health and Welfare published a *Model for a Care Programme in Suicide Prevention,* followed by other books in the field. In conjunction with the WHO "Health for All in the year 2000" programme, which has been adopted by Sweden, the target of reducing the number of suicides and suicide attempts began to attract public attention. At the follow-up consultation in Szeged 1989 it was observed that the incidence of suicide continued to increase in most countries. It was concluded that the time had come to abandon the view of suicide as an isolated problem. It would be more appropriate to perceive suicidal behaviour as a public health problem closely related to other problems such as mental illness, abuse, accidents and violence. Governments were recommended to adopt national programmes for suicide prevention based on a closer cooperation between suicidologists and other professionals in the public health field (epidemiologists, health educators etc). The Norwegian and Finnish programmes for suicide prevention inspired our Swedish efforts.

National Council

A National Council for Suicide Prevention with the task of creating a Swedish programme started its work in February 1994. The Director-generals of the National Board of Health and Welfare and the National Institute of Public Health chair the National Council, which includes representatives of the Swedish Church (the Archbishop), the Health Care Board of the Swedish Defence Forces, the National Police Board, the Federation of County Councils and the Swedish Association of Local Authorities. Professor Marie Asberg represents the Swedish Psychiatric Association. A leading medical journalist is also associated with the group.

Professor Danuta Wasserman and Information secretary Inga-Lill Ramberg (secretary of the National Council), both from the National Centre for Suicide Research and Prevention, professor Charli Eriksson from the National Institute of Public Health and associate professor Jan Beskow, injury researcher professor Jan Thorson constitute the National Council's working group.

Aims

The Council has formulated the following general aims:

1. To attain a permanent decrease in the number of suicides and suicide attempts in Sweden
2. To eliminate, as far as possible, circumstances that may result in children and young people taking their own lives.
3. To detect rising trends of suicide and suicide attempts in risk groups as early as possible.
4. To raise the general level of knowledge concerning suicidal behaviour, so that human fellowship and social measures can support people contemplating suicide or experiencing suicide or suicide attempts in their next of kin.

Principles

Pondering over death and suicide is a natural part of the struggle for life. However, in facing overwhelming problems, sudden losses, mental disorders and abuse, normal adaptation mechanisms may be insufficient and a mental crisis may be triggered. Some of these cases may be complicated by serious suicidal considerations or acts.

The task for a comprehensive suicide prevention programme is basically to increase the general public's awareness of the role of suicidal thoughts and acts in suicidal crises, in mental disorders and in other problem situations and to promote attitudes and techniques that can prevent suicide. Both modern suicidological research and cultural attitudes to suicide and suicide prevention must be considered.

Prevention Model

Suicide prevention will be developed on three levels

1. General suicide prevention. This level includes psychological, pedagogical and social support aimed at raising the general public's level of competency in handling life crises and suicidal problems. This also includes prevention and mitigation of injuries caused by suicide attempts.
2. Indirect suicide prevention. Efforts aimed at decreasing the number of suicidal acts through actions directed at predisposing factors in risk groups and in risk situations. This also includes general actions in order to delimit the access and availability of suicidal means
3. Direct suicide prevention. Efforts aimed at decreasing the number of suicidal acts among suicidal persons. This includes actions to delimit the access and availability of suicidal means on an individual basis.

Increased Awareness about Suicide and Suicide Prevention

Suicide prevention work involves the spread of knowledge and appropriate attitudes. This encompasses suicidal behaviour not only in psychic crises and in mental disorders, but also its role in the philosophy of life. First, current knowledge must be disseminated on a broad basis. In the long run it is necessary to develop interdisciplinary suicide research including aspects of epidemiology, medicine, natural science, behaviour sciences and social sciences. This must also include humanistic disciplines, as well as the study of religion. Stimulating the development of a more differentiated language about suicidal behaviour and promoting ethical discussions are two important tasks.

Better Suicide Prevention Directed Towards Risk Persons and Risk Situations

A prerequisite for suicide prevention is comprehensive knowledge about risk groups, such as people with mental disorder and substance abuse, and about risk situations that may elicit suicidal acts, such as losses and violations. A broader awareness of different forms of suicidal communication is also crucial.

Increased Professional Competence in Analyzing and Treating Suicidal Persons

Persons who work professionally with health services, rescue services, social services, religious work as well as those who work with children and youth may be concerned with suicidal problems and prevention. They may be able to support suicidal persons, either directly or indirectly via family members or peers. They need basic training in understanding and meeting suicidal people or persons with mental disorders. They must also be acquainted with available sources of help.

Interdisciplinary and Intersectorial Co-operation

An effective suicide prevention requires that suicidal problems are not perceived as isolated problems but are seen in relation to adjacent problem areas, such as questions concerning the philosophy of life, mental disorder, abuse, social problems, accidents and violence. This requires differentiated but coordinated measures by many authorities and non-profit organizations as well as religious congregations.

Systematic Evaluation

A goal-directed approach implies evaluation, follow-up and a continuous effort to secure quality. Evaluation methods must be developed.

Strategies

Enhancing Public Awareness of Suicidal Behaviour

This means striving in the long run for a thoroughly worked out and generally accepted approach to suicidal actions based on both Swedish cultural tradition and modern research.

A capacity to talk about death and to understand life-death dialogues enriches the knowledge and feeling of life. It is also a prerequisite for talking about suicide. It may also increase the capacity to handle stressful life problems in one's own life or in the life of one's fellow-beings.

Everyone needs basic knowledge about suicidal behaviour, crisis reactions and mental disorders. Everyone also needs to know that suicide is not a firm destiny but is often influenced by chance, and can be prevented. Differentiated forms of help are available.

Media presentations of this emotionally loaded but underdeveloped topic, must be presented cautiously. This must not prevent a broader, deeper public dialogue about suicidal problems.

Reports with basic data about suicide and suicide prevention as well as information material for different groups must be developed. The effect of media presentations about suicidal behaviour must be evaluated.

Support and Treatment

Suicide prevention often develops in the interaction between two persons, often in the family or at the workplace. Both persons need help and support. Such help can be given by other intimates, crisis centres, social services etc.

Treatment and care within health services should be conducted with reference to humanistic, psychological, biological and social aspects. This means early identification of suicide-prone persons, adequate treatment and follow-up but also guidance and support to intimates. "Suicide teams" following the development of knowledge, supporting other personnel groups and working out better routines are recommended as well as care programs for clinics, hospitals and regions. Further training is necessary for many categories in identification and treatment of depression and suicidal risk patients, especially at the primary care level. Efforts to enhance the quality of the care and treatment of suicidal persons at psychiatric clinics are being developed in co-operation with the Swedish Psychiatric Association.

Children and Young People

During their upbringing and schooling, children and young people need to learn to meet crises and conflicts and to master their depressive and suicidal tendencies. Suicidal communication and risk factors must be observed among pupils and students, especially when a suicidal act has occurred. Transitions between different school levels or from school to university may be especially stressful. Furthermore chaotic and destructive families must be identified and supported.

Adults

The social tendencies that further individualization and rapid change challenge our needs of satisfaction, continuity and security. Each individual needs a capacity to differentiate between problems generated on a personal and on a social level. Efforts must be focused on family problems, access to work, the psychosocial working environment and the capacity to experience connection and meaning. Attention must be paid to personal and group development. The techniques of debriefing after stressful group experiences must be applied more generally.

Elderly People

The possibility of an active and meaningful life has increased for many elderly people during the last decades. The life quality of the elderly, especially men, may now be threatened by a decreasing economy and overloaded services. The fear of disease, decreased autonomy, lack of care during the last period of life and a painful death may contribute to suicidal behaviour. Competence to handle psychic crises, depression and suicidality in higher ages as well as better care for terminally ill must be developed. In particular, the situation of elderly immigrants must be observed.

Other Risk Groups

Certain risk groups, such as immigrants and refugees, abusers of alcohol and drugs, persons infected by HIV or AIDS and persons exposed to various forms of violations, such as sexual violence or discharge under humiliating circumstances, must be especially considered.

Teaching and Training Programmes

Teaching programmes for school pupils should integrate discussions on suicidal behaviour in more general topics such as mental health, handling crises and conflicts and peer support. General training programmes for all people who may come in contact with suicidal persons must be complemented with special programs for personnel in social and mental health organizations. Training of suicide prevention teachers is a keystone in the implementation work. Methodological development of the pedagogy of teaching and training in suicidology and suicide prevention is necessary.

Reduced Availability of Suicide Means

Passive preventive methods, that is methods not requiring a new decision in every new situation, have been most successful in reducing injuries in working life and in traffic. Examples include methods that decrease the exposition to injuries and the reduction of the extent of the injury when it occurs. A general reduction of the risk for traffic deaths and in carbon monoxide intoxication would thus also reduce the number of suicidal acts with these methods.

National Expertise

The National Centre for Suicide Research and Prevention includes sections for a) information and primary prevention, b) training courses and secondary prevention and c) socio-epidemiology. The research conducted at the Centre is interdisciplinary, emphasizing psychodynamic, social and biological aspects. The centre is currently developing a surveillance system for determined and undetermined suicides in the Nordic region and in Europe, a database on current suicidological research and a national library of suicidology. The Centre must be further developed to be able to meet new demands. Interdisciplinary suicidological research at other research institutes should also generally be supported.

Systems of Regulations

A number of laws and ordinances with a bearing on suicide prevention will be scrutinized.

Implementation

The National programme was published in the autumn 1995 and then presented to the government. The National Board of Health and Welfare, the National Public Health Institute and the National Centre for Suicide Research and Prevention will be responsible for the implementation. The National Council for Suicide Prevention will function as a reference group.

The next task will be to develop direct contact with organizations and groups at various levels, stimulating them to start projects in suicide prevention. First we will approach organizations with previous experience of suicide prevention such as hot-lines, crisis centres and organizations supporting survivors after suicide as well as psychiatric patient organizations.

The programme will successively be supplemented by books, articles and manuals geared at various recipients. One booklet about existential problems entitled "Suicide: freedom and compulsion" is published in Swedish and a manuscript with brief instructions in suicide prevention aimed for the general public has been prepared. Gender aspects will be taken into account in all applicable areas. All undertakings will be systematically evaluated.

Current information on the national programme may be obtained from the National Centre for Suicide Research and Prevention, IPM, Box 230,S-171 77 Stockholm, Sweden.

Acknowledgements

This work was supported by the Swedish Council for Planning and Coordination of Research, the Swedish Medical Research Council and the Folksam Scientific Council.

References

Jansson A. Mörda för att få dö. In Jarrik A & Soderberg J. Människovärdet och makten. Om civilseringsprocessen i Stockholm 1600–1850. Summary in English. Stockholmsmonografier, Vol 118 Kommitten for Stockholmsforskning, Stockholmia forlag, 1994.

Söderberg J. Den moderna människans uppkomst. Folkets historia, 1993; 21(4): 32–33, 40–52.

Pabst Battin M. Ethical Issues in Suicide. Prentice-Hall Series in the philosophy of medicine, Prentice Hall Inc., Englewood Cliffs, New Jersey 07632, 1982.

25. England's Policy on Severe Mental Illness

RACHEL JENKINS

Introduction

Although different countries have different systems of health care, it is nonetheless generally true that countries share similar objectives for people with mental illness within a range of social and economic constraints. They therefore face difficult choices in the allocation of scarce resources between mental health and physical health and, within mental health, between the different client groups.

The Department of Health's objectives for mental illness were enunciated in the Public Health Information Strategy (1991), and have subsequently been embodied in the Health of the Nation Strategy (1992, 1993, 1994). They are firstly, to reduce the incidence and prevalence of mental disorder; secondly to reduce the mortality (both from suicide and from deaths from physical illness) associated with mental disorder; thirdly, to reduce the extent and severity of other problems associated with mental disorders, for example, poor physical health, impaired physical and social functioning, poor social circumstances, family burden; fourthly, to ensure that appropriate services and interventions are provided; fifthly to reverse the public's negative perception of mental illness by countering fear, ignorance and stigma, creating more positive social climates in which people are encouraged rather than deterred from seeking help and improving the quality of life for people with mental health problems; and sixthly, to research the causes, consequences and care of specific mental disorders.

Historical Background to Policy

Historically, large asylums were built in England during the nineteenth and early part of the twentieth century. These asylums were seen at the time as a great humanitarian step forward to solving the problems of people neglected on the streets, and delivering care, treatment, food and shelter, and occupational activities. The last large asylum was built in England in 1941, although

D. De Leo, A. Schmidtke and R.F.W. Diekstra (eds.). Suicide Prevention, 235–243.

the total number of beds in institutions did not peak until 1955. However, at the same time, a small number of pioneering psychiatrists were recognizing the damage that institutionalization can do, and had started to develop and try out methods of caring for people in the community rather than in institutions. By 1959, the Mental Health Act was firmly advocating care in the community where possible.

A number of factors have contributed to the decline in bed numbers in England. At first, in the 1940s, 50s and 60s, there were published a number of sociological critiques of the asylums, which influenced a change in social values about caring for people with mental illness. These cultural changes accompanied a greater optimism about the outcome of mental illness as more effective medical treatments were developed.

In 1961, the Minister for Health announced the asylum closure programme, and in 1975, the government published a White Paper "Better Services for the Mentally Ill" which set out the policy framework for service development, proposing local comprehensive care, including small acute units, residential accommodation and community services.

However, implementation has been slow, and expenditure on community services is still too low relative to expenditure on hospital services.

In 1986 the Audit Commission reviewed community care, and pointed out that reprovision of local comprehensive services was still far too patchy. In 1989 the Government produced two white papers, "Caring for People: Community Care in the Next Decade and Beyond" and "Working for Patients". These were consolidated into the 1990 NHS and Community Care Acts. These reforms encouraged a wider debate about the criteria for allocation of resources, and increased the need for better information on costs and cost effectiveness. In 1994 the Audit Commission reviewed the care of people with severe mental illness and again pointed out that reprovision of local comprehensive services was still inadequate.

Health of the Nation

In 1992 the Government published its strategy for "The Health of the Nation" which set out a strategic framework for the achievement of health gain in five areas, including mental illness and set targets for prevention of morbidity and mortality. So if policy is about where and how people with severe mental illness are cared for, it is also about with what aims in view, and the Health of the Nation strategy sets such aims and goals.

The Health of the Nation targets are to improve morbidity, and to reduce mortality, both in the whole population and specifically in people with a severe mental illness.

The framework to achieve the Health of the Nation targets has three dimensions, improving information and understanding, the development of local comprehensive services, and the development of good practice.

Improving Information and Understanding

(a) *National Survey of Morbidity*
The Department of Health has commissioned a national survey of psychiatric morbidity (Jenkins and Meltzer 1994, OPCS 1994, 1995).

This survey has four components, a household survey, a survey of institutions, a survey of the homeless and roofless and a supplementary sample of people with psychosis. It is producing data on population prevalence, associated risk factors, an assessment of use of services and how far needs for care are being met, and will underpin prevention strategies. Besides being the first such national survey in GB, it is the first in the world to produce simultaneous probability samples of household, institutional and homeless populations on prevalence and severity of illness, comorbidity, risk factors, social disability and use of services.

(b) *Public Information Strategy*
The government has also undertaken a three year, public information strategy costing $300,000 a year to improve public attitudes to mental illness, promote a better understanding of mental illness and of mental health services.

(c) *Monitoring of health outcomes*
The government is planning the routine measurement of health outcomes as well as of mortality (see below).

Developing Local Comprehensive Services Inputs

It is helpful to consider policy in a systems framework, of inputs, processes and outcomes. The inputs include the buildings and the people and their training, the processes refer to the activity in a service to care for patients, and the outcomes refer to the eventual health and social functioning of the patients.

These inputs are not alternatives, but are all necessary components of a whole service if it is to function well. If one component is missing, then undue pressure is placed on other components which are then prevented from operating properly. For example, in England, many districts have not yet developed the complement of new long stay provision (24 hour nursed beds) that is necessary to care for the small numbers of patients who, with the best of rehabilitation, continue to need 24 hour nursing care for many years. In the absence of such provision, these patients accumulate in acute beds, with the result that many patients needing urgent brief admissions for relapses are unable to gain such admission.

Development of Good Practice (Processes)

Mental health information systems, at their simplest, comprise a list of all those in contact with the secondary care services, by age, sex and diagnosis; but many are now far more complex, encompassing sensible core clinical information, and are hence developing clinical as well as administrative uses.

Good mental health information systems are essential for the care of severe mental illness as information is integral to assessing needs; resource management and planning; joint working between health and social care professionals; ensuring the effective delivery of appropriate care, measuring the effectiveness of different treatments, and different settings, for clinical audit and research; providing the basis for more refined contracting; and for assessing costs.

The cost information required relates to inputs, processes and outcomes. We need to know the direct expenditure levels and opportunity costs of providing structures and teams; the relative opportunity costs of deploying these structures and professionals in various ways, e.g. implementing the CPA and SR; and the costs of achieving certain specified health outcomes and the resource consequences of inadequate interventions.

The Care Programme Approach is a systematic approach to everyone in contact with secondary care services. It was introduced in the NHS in April 1991, and provider units are required in collaboration with social services departments, to make individually tailored care programmes for everyone in contact with secondary care services. The CPA calls for a systematic assessment of the health and social care needs of the patient, and for a package of care to be assembled to meet those needs, drawn up in agreement with the multidisciplinary health team, social services, the general practitioner, user and carers. A key worker should be appointed for each patient to keep close contact with the patient. There should be regular review and monitoring of the patient's needs and progress, and of the delivery of the care programme.

The key worker has a number of responsibilities, including using their professional skills collaboratively in assisting patients and maintaining regular contact with them and their carers; providing support and care in a positive assertive manner as acceptable to the patient as possible; acting as a consistent point of contact for users, carers, local authority care managers and other professionals; ensuring that the user is registered with a GP, and working in close contact with the primary health care team; and to assist in planning and monitoring the agreed care package, recording decisions and ensuring regular review.

Supervision Registers were introduced in 1994 to identify patients who are at significant risk of suicide, serious harm to others or serious self neglect. They are a priority list of those on the CPA who are most at risk, and will be incorporated into the development of mental health information systems.

Routine measurement of mental health outcomes

Jenkins (1990) argued that processes, while important in their own right, are not adequate proxies for outcomes, and that the best method for measuring outcomes is to use direct measurement of health and social functioning. This has long been possible in a research context, using instruments like the Present State Examination for psychosis or the Clinical Interview Schedule for neurosis. But these instruments are too lengthy for use in a routine, clinical setting, and so, in order to measure our first target in the health of the nation, the government commissioned the Royal College of Psychiatrists Research Unit to develop a brief scale for use as an integral part of routine CPA reviews. This instrument (HoNOS) has undergone satisfactory field trials and will be introduced in 1996/97.

Besides monitoring the first target in Health of the Nation, HoNOS provides standardized measures of caseload severity for managing the clinical caseload, provides measures of variations for clinical audit, and provides information for casemix classification and delineation of Health Care Resource Groups. The data will be collected within the context of a core clinical minimum data set.

Purchaser-Provider Split

England has a national health service, with publicly funded health and social care. The Department of Health allocates revenue to regional health authorities by a formula using resident population figures, age and mortality data (as a proxy for morbidity). RHAs allocate resources to district health authorities and GP fundholders who are the purchasers in the new internal market. Purchasers then decide what to purchase in the light of local needs, local priorities and central guidelines.

The purchaser-provider split was introduced into the NHS by the 1990 NHS and Community Care Act in order to inject market forces into the NHS, to improve decision making about the deployment of resources, making them purchaser rather than provider, hence needs led rather than supply led, and to improve service innovation and quality.

The contracting process is the main mechanism for purchasers to improve care, and contracts should therefore focus on the needs of patients, should include health gain targets, should specify clear service objectives and should provide incentives and schetions. In order to do this, purchasers need to consult with patients and clinicians. assess local needs, improve information systems, use the results of local clinical audits, and engage in open and explicit negotiations with providers.

To support purchasers in these difficult tasks, the Department of Health has produced a Mental Illness Key Area handbook (DH 1993) which has now been revised (DH 1994) and has carried out the national survey of psychiatric morbidity (OPCS 1994,1995 Jenkins and Meltzer) described earlier. Furthermore, the Health of the Nation Outcome scales will soon be ready for use in the contracting process.

Policy on minor and moderate mental illness

Effective policy on severe mental illness requires effective policy on minor and moderate mental illness and effective policy on mentally disordered offenders, i.e. effective policies at both the upper and lower boundaries of severity, otherwise the care of people with severe mental illness is compromised by the pressures from the small minority of mentally disordered offenders, and by the pressures from the overwhelming majority of people with minor and moderate mental illness. Most people with minor or moderate mental illness consult their GP during the course of the year and so primary care strategies are an important component of mental health policy. These strategies include early detection and treatment, screening of the high risk groups, better methods of deploying the primary care team and better criteria for referral to secondary care.

The Department of Health has commissioned a number of developmental projects to improve mental health care in general practice, and these are being evaluated in terms of cost effectiveness, change in GP knowledge and attitudes, and patient outcomes. The Department of Health has also established a Senior GP Fellow, at the Royal College of GPs, to take a national lead in developing the continuing education of GPs about mental health, and a Senior Primary Care Nurse Facilitator to take a national lead in developing the continuing education of primary care facilities about mental health. Both of these national figures have identified an assistant in each Region in the country to help them prepare teaching materials and cascade them.

The Government is also tackling mental health in the workplace, where again the principal issues are the same as for primary care, i.e. there is a high prevalence of morbidity, a substantial proportion is undetected, "hidden" morbidity, and that which is recognized is not always managed optimally. This untreated and poorly managed morbidity has vast public health consequences (the CBI has recently estimated the cost at $5 billion to the economy) both for the economy, the individual and their families. The Department of Health has therefore convened an interagency group to co-ordinate activities on mental health in the workplace, comprising membership from CBI, TUC, Department of Employment, Health and Safety Executive, ACAS, Health Education Authority, and Institute of Personnel Managers. The Department has organized conferences (Jenkins and Coney 1991, Jenkins and Warman 1993), leaflets for employers (Department of Health 93, 94) and stands at national CBI and IPM conferences to raise awareness amongst employers of the importance of a mentally as well as physically healthy workforce. Workplace strategies need to be comprehensive rather than piecemeal in design, and need to value the mental health of the workforce, promote understanding and reduce stigma, reduce workplace stress, improve the prompt detection and management of illness at work, improve rehabilitation back to work and encourage the development of workplace health policies that address mental as well as physical health.

Suicide Prevention

Suicide reduction requires a multipronged approach, including improved management of depression in general practice, education of primary and secondary health and social care professionals about assessment and management of suicidal risk, reducing the availability of certain methods used in suicide, targeting high risk groups, developing mental health services, audit and research.

Improved management of depression in general practice

Recognition by GPs of depression occurs in only about 50%. Improvement in recognition and management leads to better outcomes and also to improvements in suicide rates. Therefore the Defeat Depression campaign, organized by the Royal Colleges of General Practitioners and Psychiatrists, has developed training packages for GPs and Management Guidelines, produced and distributed by the Department of Health. The Department of Health's funding of a senior GP Fellow to work with regional advisors in mental health in primary care, to produce educational materials for course organizers and GP tutors, which was referred to earlier in this paper is also helpful in this respect. Furthermore the Department of Health has funded a number of development projects to improve assessment, diagnosis, management and treatment of primary care, and auditing the care of depression.

Reducing access to the means of suicide

We know, from the coal gas and barbiturate experiences, that removing access to a method of suicide only results in partial substitution by other methods. It is therefore worthwhile to explore the possibilities of reducing the availability of certain methods of suicide e.g. by increased use of catalytic converters on vehicle exhausts reducing toxicity of fumes, adaptation of car exhausts to prevent fitting of tubes; the use of blister packs for paracetamol, or marketing paracetamol combined with the antidote, better precautionary measures in relation to supervision of guns. (Approximately one farmer a fortnight in England has killed himself with a gun for the last ten years).

Targeting high risk groups

It is especially helpful to target appropriate supportive measures and early detection of suicidal risk at those groups who are at particular high risk of suicide. These include people in prison (appropriate measures include improving reception procedures, the use of buddies and the Samaritans, and the training of prison medical staff and prison officers); some high risk occupational groups including doctors, nurses, vets, pharmacists, dentists and farmers (the Samaritans are working with farmers organizations); people who have recently attempted suicide (by improving assessment and management in Accident and Emergency Department); and of course people with Severe Mental Illness.

Developing mental health services

Developing mental health services can be expected to assist in reducing the suicide rate of people with severe mental illness by producing local, accessible services with effective supervision systems, by measures to improve the support and supervision of users of services in the community, especially the care programme approach, care management and the development of supervision registers and other effective information systems. These assist in ensuring that when suicidal risk or other risk emerges, the person is in contact with a key worker who they know, can talk to, and who can ensure that appropriate reassessment and management occurs. Clear, agreed observation policies are necessary, for managing suicidal people in hospital and for managing those who have deliberately self harmed.

Audit and Research

Audit can improve practice. The Government has set up a National Confidential Inquiry into Homicides and Suicides by Mentally Ill People which will provide a continuing source of information to assist prevention. Services are also encouraged to set up local multidisciplinary audits of suicide and many have done so. The Government is also investing in a series of detailed research studies of suicide in order to establish better methods of predicting those at greatest risk and methods of prevention.

Conclusions

England has set an evidence based mental health strategy which encompasses where and how people are cared for and with what aims and goals in view. It has ensured that policy is well rooted in the known epidemiology of mental disorders, that policy does not only focus on those with severe mental disorders who need specific care, but also on those with less severe disorders in primary care, in the workplace and in prisons, and also on the integration and interface of mental health services with other agencies. It has also rooked policy within a coherent framework of prevention mental health promotion (primary, secondary and tertiary prevention and prevention of mortality), and is driving that policy by measuring health outcomes as well as inputs and processes.

MIX
Papier aus verantwortungsvollen Quellen
Paper from responsible sources
FSC® C105338

If you have any concerns about our products,
you can contact us on
ProductSafety@springernature.com

In case Publisher is established outside the EU,
the EU authorized representative is:
Springer Nature Customer Service Center GmbH
Europaplatz 3, 69115 Heidelberg, Germany

Printed by Libri Plureos GmbH
in Hamburg, Germany